T0320811

Spaces of Enlightenment Science

Knowledge Infrastructure and Knowledge Economy

Edited by

Karel Davids (*VU University, Amsterdam*)
Larry Stewart (*University of Saskatchewan, Saskatoon*)

VOLUME 8

Spaces of
Enlightenment Science

Edited by

Gordon McOuat and Larry Stewart

BRILL

LEIDEN | BOSTON

Support for the project has been provided by the Social Sciences and Humanities Research Council of Canada and its Strategic Cluster Project, "Situating Science": www.situsci.ca

Cover illustration: Hindostan, by Thomas Luny, c. 1793. Permission of the National Maritime Museum, Greenwich (also included as figure 8.2).

The Library of Congress Cataloging-in-Publication Data is available online at https://catalog.loc.gov
LC record available at https://lccn.loc.gov/2021051175

Typeface for the Latin, Greek, and Cyrillic scripts: "Brill". See and download: brill.com/brill-typeface.

ISSN 1877-2323
ISBN 978-90-04-50121-8 (hardback)
ISBN 978-90-04-50122-5 (e-book)

This book is printed on acid-free paper and produced in a sustainable manner.

Contents

Figures

Notes on Contributors

Victor D. Boantza

is a professor in the Program for the History of Science, Technology, and Medicine at the University of Minnesota. He is working on the history of the early modern physical sciences, especially matter theory, chemistry, and experimental science.

Margaret Carlyle

is an Assistant Professor at the University of British Columbia Okanagan. She is an historian of science, medicine, and technology with specialization in early modern France in both European and global contexts. Her current research focuses on women's role in the formation of scientific knowledge—as experimentalists, artisans, and translators. She engages with archival and museum sources, including manuscripts, printed texts, drawings, engravings, and objects. Themes of interest include: the visual culture of science; medical techniques and technologies; gender and science.

Rob Iliffe

is Professor of History of Science at the University of Oxford. He has written widely on the history of early modern science, and is the author of *A Very Short Introduction to Newton* (Oxford University Press, 2007) and *Priest of Nature: the Religious Worlds of Isaac Newton* (Oxford University Press, 2017). He is the co-director of the Newton Project, and the director of the Newton Mint Papers Project.

Jasmine Kilburn-Toppin

joined Cardiff University as Lecturer in Early Modern History in January 2020, with particular research interests in artisanal cultures, urban space and architecture, and networks of craft and 'scientific' knowledge in Early Modern England. Her monograph, *Crafting identities: artisan culture in London, c.1550–1640* (Manchester University Press, forthcoming 2021), argues that the livery halls of artisanal guilds became multifunctional sites for technical innovation, civic memorialisation, and social and political exchange.

Trevor H. Levere

is a historian of science who has written mainly about the history of chemistry, and also about science and the Canadian Arctic; science and Romanticism; scientific instruments; and the history of science in Canada. His books include *Affinity and Matter* (Oxford University Press, 1971), *Poetry Realized in*

Nature (Cambridge University Press, 1981), *Science and the Canadian Arctic* (Cambridge University Press, 1993), and *Transforming Matter: a History of Chemistry from Alchemy to the Buckyball* (Johns Hopkins University Press, 2001). He is currently exploring the interplay between concepts, apparatus and experimentation in the history of chemistry. He is University Professor Emeritus at the University of Toronto, where he was first appointed in 1968.

Alice Marples
is a historian of science and medicine in Britain and its colonial networks c.1650–1850, currently working on the Newton Project at the University of Oxford. Her work focuses on the collection and management of manuscript, material, and human resources across scientific institutions and international networks. Her first book, *The Transactioneer: Hans Sloane and the Rise of Public Natural History in Eighteenth-Century Britain,* is forthcoming with Johns Hopkins University Press.

Gordon McOuat
teaches in the History of Science and Technology Program at the University of King's College/Dalhousie University. His work focuses on the history and philosophy of logic, classification, and the origins of "natural kinds", and, globally, the cross boundry circulations of knowledge.

Larry Stewart
is Professor Emeritus at the University of Saskatchewan and visiting scholar at the University of King's College, Halifax, Canada. He has published numerous works on the public response to science in the early-modern world and is currently writing a study of the scientific life of the itinerant lecturer James Dinwiddie.

Marie Thébaud-Sorger
is Researcher (with tenure) at the French National Center for Scientific Research (CNRS) based at the Centre Alexandre-Koyré, Paris, and associated to the Maison français in Oxford where she ran a History of Science programme (2017–2020) and is co-leading the Interdisciplinary Network, *Writing technologies* at the Oxford Research Centre in the Humanities (TORCH). She published extensively on the lighter-than-air machines in French and European societies in her first monograph, "Ballooning in the Age of Enlightenment" (*L'aérostation au temps des Lumières* (Presses Universitaires de Rennes, 2009), the culture of flight, on the public sphere of technology, and how inventive practices in Eighteenth Century Europe shaped communities around technical improve-

ments, especially focusing on the management of fires and "stale/noxious/ foul air". Recent work includes "Changing Scale to Master Nature" (2020), and "Capturing the Invisible: Heat Steam and Gases at the Crossroads of Inventive Practices (2017).

   Simon Werrett
is Professor of the History of Science in the Department of Science and Technology Studies at University College London. Werrett's work explores exchanges between practical arts and sciences in the sixteenth to nineteenth centuries. His first monograph *Fireworks: Pyrotechnic Arts and Sciences in European History* (University of Chicago Press, 2010) examines fireworks and natural philosophy in Britain, France and Russia c.1500–1800. His second book *Thrifty Science: Making the Most of Materials in the History of Experiment* (University of Chicago Press, 2019) explores the history of experimenting in the home. He has also co-edited, with Lissa Roberts, *Compound Histories: Materials, Production, and Governance*, 1760–1840 (Brill, 2017).

The editors would like to acknowledge the work of our indefatigable research assistant, Sophie Lawall, for bringing this text to completion.

CHAPTER 1

Introduction

Gordon McOuat and Larry Stewart

1   The Importance of Place

Where did we do Enlightenment Science? The places of science, its movements, its locales have often proved problematic, defined in its essence by the
scientific activities which took place within often regal and aristocratic institutions.[1] It cannot go unnoticed that philosophic enlightenment, in practise
and in places, could be seen to have underpinned some of the dramatic social
and political changes of the 17th and 18th centuries. Indeed those, like Edmund
Burke, opposed to enlightenment thinking and to its apparent revolutionary consequences, believed the 'calculators' of the French Revolution would
grind grand institutions and established constitutions to dust. Burke famously
declared, in 1790, "The Age of Chivalry is gone. That of sophisters, economists,
and calculators has succeeded; and the glory of Europe is extinguished for
ever."[2] Burke's revulsion at the stirrings of revolution was, in some measure, a
reaction to the failure of French institutions but even more broadly to fears for
monarchy and a hereditary constitution in general. Brilliantly written, Burke
blasted the "mazes of metaphysic sophistry," the promoters of "this barbarous
philosophy, which is the offspring of cold hearts and muddy understandings...,"
as if a constitution "be a problem of arithmetic."[3] Institutions mattered and
they did have consequences. This was surely one version of conspiracy thinking that energized those who would sound the tocsin over secret actors seeking
to upset an established order.[4]

---

1   Cf. Fokko Jan Dijksterhuis and Andreas Weber, "The Netherlands as a Laboratory of Know-
    ing," introduction to Dijksterhuis, Weber and Huib J. Zuidervaart, *Locations of Knowledge in
    Dutch Contexts* (Leiden: Brill, 2019), 1–14, at 8–9.
2   Edmund Burke, *Reflections on the Revolution in France. A Critical Edition*, ed. J.C.D. Clark
    (Stanford: Stanford University Press, 2001), 238.
3   Burke, *Reflections*, 169, 240, 209.
4   See the Edinburgh professor John Robison and his exposure of the illuminati in the much-
    republished *Proofs of a Conspiracy against all the Religions and Governments of Europe*,

In the essays which follow we explore scientific places and spaces within an emerging Enlightenment. Notably, more than thirty years ago Maurice Crosland revealed the ways in which politics contaminated the debate over Joseph Priestley's contribution to the chemistry of airs.[5] An underlying theme is the role science played in numerous spaces in urban Europe and through the reach of Empire. As Adi Ophir and Steven Shapin have put it, "the conditions of our knowledge vary according to our placement in social and physical space."[6] The boundaries of any institution define the limits of local knowledge just as scientific notions can reassert the significance of spaces themselves— as in botanic gardens, in libraries and laboratories, in museums, even in the reach of empires. Thus, we can recognize that influential establishments like the Royal Society of London or the Académie des Sciences in Paris both convincingly asserted a right to adjudicate the latest discoveries, to deliver their imprimatur to publications and promote the dissemination of scientific news. We might well ask whether these induced, by virtue of royal favour, "a quintessential theatre of a constructed nature" that would serve the use of monarch and state. Seizing the mantel of expertise was undoubtedly a means to assert authority,[7] but such institutions, variously tied to royalty and empire, were far from alone.

---

carried on in the Secret Meetings of Free-Masons, Illuminati and Reading Societies, etc., Collected from Good Authorities (Edinburgh: Creech, Cadell and Davies, 1797). Cf. Michael Taylor, "British Conservatism, the Illuminati and the Conspiracy Theory of the French Revolution, 1797–1802," *Eighteenth-Century Studies* 47 (Spring, 2014): 293–212; and Mike Jay, "Darkness over All" (2020), https://mikejay.net/darkness-over-all/

5   Maurice Crosland, "The Image of Science as a Threat: Burke versus Priestley and the 'Philosophic Revolution'," *British Journal for the History of Science* 20 (July, 1987): 277–307.

6   Adi Ophir and Steven Shapin, "The Place of Knowledge: A Methodological Survey," *Science in Context* 4 (1991): 3–21, at p. 9. Cf. William Clark, Jan Golinksi, and Simon Schaffer, eds., *The Sciences in Enlightened Europe* (Chicago and London: University of Chicago Press, 1999).

7   Stephan Van Damme, "The Academization of Parisian Science (1660–1789): Review Essay on a Spatial Turn," in Mordechai Feingold and Guilia Giannini, eds., *The Institutionalization of Science in Early Modern Europe* (Amsterdam: Brill, 2019), 20–51, at 38, 45–46. See Lissa Roberts, "Going Dutch: Situating Science in the Dutch Enlightenment", in Clark, Golinski, and Schaffer, *The Sciences in Enlightened Europe,* 350–388; Van Damme, "'The World is Too Large': Philosophical Mobility and Urban Space in Seventeenth- and Eighteenth-Century Paris," *French Historical Studies* 29 (Summer, 2006): 370–406; Fokko Jan Dijksterhuis and Andreas Weber, "The Netherlands as a Laboratory of Knowing," 1–14.

2   Ruminating About Place, Space, and "Universal" Science

It has been thirty years or so since "place" in science received its first significant reconsideration. [8] It did so as an integral part of the turn towards "situated science" – examinations of the practice of science and science in the making. The turn towards scientific "practice" was itself an unruly offspring of two or three movements, one in the philosophy of science and another in the sociology of science and the history of science. For its part the philosophical move engaged a recrudescent interest in scientific experiment (as opposed to "theory"), taking up the Wittgensteinian injunction to "have a look", mainly at the way in which scientific experiment and practice actually proceeded *in place*.[9] For its part, the so-called "Strong Program" in the sociology of science upped the ante, pushing aside weaker socio-historical injunctions to "contextualise science" by giving pride of place to "place", or rather, pride of local interests and actions over abstract theoretical content in the study and formation of scientific life and concepts. Historians joined in. Each of these found "place" as an important aspect of the contingencies of scientific development.

Recovering these "places" for science seemed to buck at a longstanding theoretical conviction that science, as abstract knowledge, needed no *place* in particular – or, rather, science's main ambition is to transcend the local and ascend to the universal. Science, so the story about the Enlightenment goes, works *towards* the universal, a-temporal, a-spacial, leaving its local behind.[10] Philosopher Thomas Nagel captured the received mood when he revisited

---

8   Geographers of knowledge led one of the charges. See David Livingstone "The Spaces of Knowledge: Contributions towards a Historical Geography of Science." *Environment and Planning* D 13 (1995): 5–34; Charles W.J. Withers, "Geography, Natural History and the Eighteenth-Century Enlightenment: Putting the World in its Place," *History Workshop Journal* 39 (1995): 136–163. For a survey, see Diarmid A. Finnegan, "The Spatial Turn: Geographical Approaches in the History of Science," *Journal of the History of Biology* 41 (2007): 319–388.

9   The key text here is Ian Hacking, *Representing and Intervening* (Cambridge, Mass: Harvard University Press, 1983). We say "recrudescent", for the study of experiment has always been an important aspect of the philosophy of science, albeit often taking second fiddle to "theory". See the survey in Hans Radder, "Towards a More Developed Philosophy of Scientific Experimentation," in H. Radder, ed., *The Philosophy of Scientific Explanation* (Pittsburg: University of Pittsburg Press, 2003), 1–18, updated in Hans Radder, "The Philosophy of Scientific Experimentation: A Review," *Automated Experimentation* 1 (2009): 2.

10   We recognise this move as early as the Aristotelian corpus, where Aristotle's view of substance and participation did not allow for any science of the particular, and defined place as distinct from space. On Aristotle's place, see Edward Casey, *The Fate of Place: A Philosophical History* (Berkeley: University of California Press, 1997).

the virtues of the "view from nowhere", the notion of a universal knowledge becoming "objective" by overcoming the limits of a merely local perspective.[11] Science aspires to transcend the local "prejudices", to use Edmund Burke's unfortunate term.[12] In our received view the very nature of science and its methodology settles out into spaces of local-lessness – the very idea of a universal law necessarily transcends the parochial. Science's universal application offers robustness in predictive and explanatory power anywhere, always.

Lately philosophers have been cautioning us against accepting that universality received from the story of the Enlightenment science. Nancy Cartwright, for example, wants us to acknowledge a "dappled" world, where local contingency gives us the measure and the ground of a good science.[13] As Gaston Bachelard once pointed out, the attraction of space was its connection to philosophical solitude and intellectual creativity. This, he soon asserted, calls for a realm of action along with the apparent seductions of intellect and imagination.[14] This is significant for it meant, ultimately, the recognition that the notions of an absolute, even of gravities or attractions, may well be defined differently from one location to the other. Hence, David Livingstone recently referred us to Voltaire's proposition that, within an apparent Newtonian universe, views of the natural world differed markedly between Paris and London during the early eighteenth century.[15] Even the notion of an absolute was not absolutely accepted —especially by those focussed on natural theology in the early-modern world, like Leibniz or his correspondent (and Newton's disciple) Samuel Clarke. Their differences focussed largely on the meaning of space as "sensorium". As the essays produced here help reveal, if still needed, that all space is a social entity even when contemplating the universal laws operating in natural spaces.

11   Thomas Nagel, *View from Nowhere* (Oxford: Oxford University Press, 1986).

12   Edmund Burke, *Reflections on the Revolution in France*, 179, Part 2, "In defence of prejudices." Burke was being positive about these prejudices, warning us about the news from nowhere arising out of the new forces unleashed by the 'calculators' he believed behind the French Revolution and its overt universalism.

13   Nancy Cartwright, *The Dappled World: A Study of the Boundaries of Science* (Cambridge: Cambridge University Press, 1999).

14   Gaston Bachelard, *The Poetics of Space,* ed. John R. Stilgoe, trans. Maria Jolas (Boston: Beacon Press, 1969, 1994), 10–12; Cf. his much earlier, *L'expérience de l'espace dans la physique contemporaine* (Paris: Félix Alcan, 1937).

15   Cf. David N. Livingstone, *Putting Science in Its Place. Geographies of Scientific Knowledge* (Chicago and London: University of Chicago Press, 2003), 91.

3   Local Origins: Turning Around a Needham Question

There always remained a certain irony, perhaps a rather grievous one, regarding the notion of a transcendent, local-less, view of science – an irony that has become more acute under the impact of recent post-colonial studies. It's about both origins of science and place of practice. For, even though it points towards purported universality, "science" as we now might call a certain type of knowledge and practice, happily traces its origin(s) to one particular time, one predominant place: namely, to Europe – and, at that, specific *places* in Europe – in the years of the 16th-18th centuries (and beyond).[16] The special conditions of this origin have been a focus of historians of science, from the meticulous early work of Robert Merton, following up hints from Boris Hessen and Max Weber, all the way to recent attempts to discover what it *means* to be modern in science.[17] This historiographical question asks what was "special" about the "place" of Europe that gave rise to such a universalist enterprise? What was the soil upon which such knowledge rose and flourished? Latterly, it has provided fuel for counterfactual inquiries about modernity and development. The nagging "Needham Problem" reasserts itself here: As the sage historian of Chinese science, Joseph Needham, asked: "Why Europe? Why not elsewhere?"[18] It isn't surprising that investigators would find exactly what they searching for: in Merton's case, driven by concerns of inter-war tensions, by the creation of a specific set of liberal-democratic commitments to certain "ideals" about trans-locality, the possession of knowledge as property, and the

---

16   The latest view is David Wootten, *The Invention of Science: A New History of the Scientific Revolution* (London: Alan Lane, 2015). Cf. Stephen Shapin, *The Scientific Revolution* (Chicago: University of Chicago Press, 1996); and Lissa Roberts, "Producing (in) Europe and Asia, 1750–1850 and Karel Davids, "On Machines, Self-Organization, and the Global Traveling of Knowledge, circa 1500–1900," *Isis* 106 (December, 2015): 857–865, 866–973. An earlier, Heideggerian-influenced, versions of this new world of *placelessness*, can be found in the classic Alexander Koyré, *From the Closed World to the Infinite Universe* (Baltimore: Johns Hopkins University Press, 1968).

17   Robert K. Merton, "Science, Technology and Society in Seventeenth Century England," *Osiris* 4 (1938): 360–632; Boris Hessen, "The Social and Economic Roots of Newton's Principia," in Nicolai I. Bukharin, ed. *Science at the Crossroads* (London, 1931. Reprint New York: Frank Cass, 1971), 151–212. See Bruno Latour, *Science in Action* (Cambridge, Mass.: Harvard University Press, 1988).

18   Joseph Needham, *The Grand Titration: Science and Society in East and West* (Toronto: University of Toronto Press, 1969), 16. Acres of research have been sacrificed showing that the question itself is flawed from the start (along the lines of: "why didn't Buddhism arise first in 18th century Scotland"). But the discussion asks the question, why "location"?

universal *goals* of science, the main thrust of which was "universalism" and a persistent commitment to go beyond the local.[19] This itself was a "normative" claim: history of science would remind us which types of society facilitated the full flowering of science (i.e., liberal democracy) and which ones didn't (not liberal democracy). These had particular places of origin, but pushed towards the universal. Indeed, the claim that parochial interference was inimical to scientific discovery was much championed by those who thought of the grand tour of Europe by Humphry Davy and Michael Faraday in the Napoleonic period. In other words, nations could not undermine the universal, a notion that for some historians also ultimately served a cold-war audience well.[20]

A corollary of this model involves tracing the *dissemination* of this universalism of science – the passage from centre to periphery, a spreading out through space, over time, of universalising knowledge and its associated practices across the world.[21] It is surprising how both defenders and critics of modernity and modern science often alighted on the same story – i.e., the spread of "universalism" from centre to periphery. One version celebrates the spread of Rationality against local ignorance and interests.[22] Another, less festive, more postcolonial, version laments the elimination of local knowledges along with their intimate ties to place and time.[23] Here both critics and supporters rehearsed much of the same tune – a replacement of the local by a universalising project – the displacement of *place*. But this also an example of the

---

19   Robert K. Merton, "The Normative Structure of Science," in Robert K. Merton, *The Sociology of Science: Theoretical and Empirical Investigations* (Chicago, IL: University of Chicago Press, 1979). See Joseph Ben-David, *The Scientist's Role in Society* (Englewood Cliffs, N.J.: Prentice-Hall, 1971).

20   Cf. L. Pearce Williams, *Michael Faraday. A Biography* (London: Chapman and Hall, 1965), 30ff.

21   The classic study is in G. Basalla, "The spread of Western science. A three-stage model describes the introduction of modern science into any non-European nation." *Science* 156 (1967): 611–622; Ashis Nandy, *Science, Hegemony and Violence: A Requiem for Modernity* (Oxford: Oxford University Press, 1988); Samuel Huntington, *Clash of Civilizations and the Making of the World Order* (New York: Simon and Schuster, 2001). This was the standard explanation of the so-called failure of the Macartney mission to China in the late 18th Century. See Larry Stewart, this volume. Cf. Kapil Raj, *Relocating Modern Science Circulation and the Construction of Knowledge in South Asia and Europe, 1650–1900* (New York: Palgrave, 2007) and Warwick Anderson, "Remembering the Spread of Western Science," *Historical Records of Australian Science* 29 (2018), 73.

22   The most egregious version can be found in Huntington, *Clash of Civilizations*.

23   Gayatri Spivak, *A Critique of Postcolonial Reason* (Cambridge, Mass.: Harvard University Press, 1998); Susantha Goonitalake, *Aborted Discovery* (London: Zed, 1984); cf. Chandra Mukerji, *A Fragile Power: Scientists and the State* (Princeton: Princeton University Press, 1989); Carol Harrison and Ann Johnson, eds., *National Identity: The Role of Science and Technology* (Chicago: Chicago University Press); Maureen McNeil, "Postcolonial Technoscience," *Science as Culture* 14 (2005): 105–112.

"enormous condescension of posterity" to use a phrase of the influential British historian E.P. Thompson.[24] Because we now conceive of science as essentially universal, the same in every instance in the world—thereby laying the ground of its prestige over the parochial and the political—we forget that there were many prior propositions of universalism in many places throughout the Enlightenment. One could even argue that was the essence of Enlightenment.

Working against such universalisation, social studies of scientific knowledge zeroed in on the process of science *in the making*, which is, ipso facto, *always* local. The turn to practices – the processes of science – also revealed place. Here universalization just didn't count, or it had to be "explained". The universal (if indeed there was one) was being *built*, out of local resources, ready to hand and certainly hardly placeless. Thus, space was defined by uses. This turn to the local prided itself on avoiding *any* question of transcendence, and produced, for a time, endless articulations of micro-studies, all the more centred on a particular time and place. Even the Enlightenment went local.[25] Recent focus on "circulations of knowledge" gives locality to each point of an exchange, each local limit and engagement of the movement of ideas, materials, persons, styles and objects.[26] In a sense this book follows from that move.

There are many examples on which we might draw that force us to emerge from the pretence of the universal promoted by early-modern scientific societies, whether in Paris, Berlin, London or Philadelphia. These were uniformly urban but, once we move from debating concepts of natural philosophy and their comprehension to practice and performance, of necessity we will find ourselves in many sites of science not confined to the influence of the pre-eminent and regal. In other words, authority in early-modern science was defined not solely by organization but increasingly by degrees of engagement. Interestingly, as early-modern cities created museums and sites of performance, seemingly metropolitan analogues of discoveries in the world of nature, they

24   E.P. Thompson, *The Making of the English Working Class* (New York: Vintage, 1963), 12. We thank Simon Schaffer for recently reminding us of the necessity of saving the past of science from the condescension of the present.

25   See the pioneering collection, Roy Porter and Mikuláš Teich, eds. *The Enlightenment in National Context* (Cambridge: Cambridge University Press, 1981); Cf. Mikulas Teich, "How it all Began: From the *Enlightenment in National Context* to *Revolution in History*," *History of Science* 41 (2003): 335–343.

26   See Bernard Lightman, Gordon McOuat and Larry Stewart, eds. *The Circulation of Knowledge Between Britain, India and China* (Amsterdam: Brill, 2013); Suman Seth, "Putting Knowledge in its Place: Science, Colonialism, and the Postcolonial," *Postcolonial Studies* 12 (2009), 373–388. For a comprehensive overview of the state of the field, see Johan Östling, "Circulation, Arenas, and the Quest for Public Knowledge: Historiographic Currents and Analytic Frameworks," *History and Theory* 59 (2020): 111–126.

also reimagined the local as "the effect of a series of operations..."[27] In this way the universal became local. It could appear in the library or the laboratory, in the observatory or botanic gardens, on ships and in the grasp of European empires, or in provincial clubs and local societies. There were numerous places like these such as scientific gatherings in Dijon, Nantes, or Montpellier, in the Spalding Gentleman's Society in Lincolnshire, or in small groups "down market" like London's Spitalfields Mathematical Society — or even in transitory places like coffee houses or taverns which did not restrict their gatherings by social status.[28]

4   Multiple Places

Decentring the notion of top-down or centre/periphery dissemination, we have discovered and imagined multiple "confluences" making up Enlightenment and modern science. We might say, *multiple* modernities, tracing several streams that come together (and sometimes separate) throughout the tributaries and oceans of knowledge.[29] Such tracings rely on the irreducible localism of science, even in its claims to universalism. "Cosmopolitanism", although facing

27   See Van Damme, "'The World is too Large'", 381–382, n. 13, citing the anthropologist Michèle de La Pradelle, "La ville des anthropologues," in *La ville et l'urbain* (Paris: La Decouverte, 2000).

28   Cf. *inter alia* Larry Stewart and Paul Weindling, "Philosophical Threads: Natural Philosophy and Public Experiment Among the Weavers of Spitalfields," *British Journal for the History of Science* 28 (March, 1995): 37–62; and Stewart, "Other Centres of Calculation, or, Where the Royal Society Didn't Count: Commerce, Coffee-Houses and Natural Philosophy in Early Modern London," *British Journal for the History of* Science 32 (June, 1999): 133–153; Richard Sorrenson, "The Ship as a Scientific Instrument," *Osiris* 11 (1996): 221–236; Rob Iliffe, "'Aplatisseur du Monde et de Cassini': Maupertuis, Precision Measurement, and the Shape of the Earth in the 1730s," *History of Science* 31 (December,1993): 335–375; Mary Terrall, *The Man Who Flattened the Earth. Maupertuis and the Sciences in the Enlightenment* (Chicago and London: University of Chicago Press, 2002); Marie Thébaud-Sorger, *L'aérostation au temps des Lumières* (Rennes: Presses Universitaires de Rennes, 2009); Larrie D. Ferreiro, *Measure of the Earth. The Enlightenment Expedition that Reshaped the World* (New York: Basic Books, 2011); Neil Safier, *Measuring the World. Enlightenment Science and South America* (Chicago and London, 2012); Bruno Belhoste, *Paris Savant. Capital of Science in the Age of Enlightenment*, trans. Susan Emanuel (Oxford and New York: Oxford University Press, 2019); Fokko Jan Dijksterhuis, Andreas Weber, and Huib J. Zuidervaart, eds. *Locations of Knowledge in Dutch Contexts* (Leiden and Boston: Brill, 2019); Jordan Goodman, *Planting the World. Joseph Banks and His Collectors. An Adventurous History of Botany* (London: William Collins, 2020).

29   See Arun Bala, *Dialogue of Civilizations in the Birth of Modern Science* (New York: Palgrave, 2006).

no small tribulations in our recent political climate, offers one way of seeing knowledge in transit, moving from place to place, forming a kind of pidgin.[30] But there are two types (at least) of cosmopolitanism: the universalism of the global traveller, at home everywhere and nowhere, or the more local openness to strangers, the reformulations and translations under local conditions.[31] The former turns to place and space as an annoyance, an unwanted anchor. The latter engages multiple locals, all places where knowledge gets "done", but all still open to translation, acceptance of difference, and transportability. That form of cosmopolitanism is always aware of place, and the localities of knowledge: how knowledge might be done *there* and how it was encountered across distances in its "where-ness".

We endeavour here to take matters further and go into "space" – the "whereness" where science gets done, the locatedness and limitations of space. The sense of nature is fixed by where science is conducted. "Where" provides warrant to the facts and to the practitioners of science.[32] This then claims that the practise of science is fundamentally social. The proposition of space as a fundamental realm of objectivity has perhaps induced a kind of self-deception or wish-fulfilment. Ultimately, any historical examination of space explored the social nature of space which in turn promoted activities, both of a productive and intellectual kind. Further, space could also prove limiting as in, for example, the constraints of dimensions of laboratories or observatories, or of equipment that may be available to an observer. As Henri Lefebvre put it, "Where natural space exists, and even more so where social space exists, the movement from obscurity to enlightenment—the process of decipherment—is perpetual."[33] The "where" of science is the space of practise and, thus, is social in the sense of knowledge production and grounding of credibility. Space defines the activity that may occur; it is the broader interpretation of the nature of nature that comes later. Space is thus "produced before being

---

30 For the rise of "cosmopolitanism" in the early modern period see Margaret Jacob, *Strangers Nowhere in the World: The Rise of Cosmopolitanism in Early Modern Europe* (Philadelphia, University of Pennsylvania Press, 2006). For vernacular cosmopolitanism, see Ananta Kumar Giri, ed., *Beyond Cosmopolitanism* (New York: Palgrave, 2018).

31 On "cosmopolitan contamination", see Kwame Anthony Appiah, *Cosmopolitanism* (New York: W.W. Norton, 2006). Cf. Margaret C. Jacob, *The Secular Enlightenment* (Princeton: Princeton University Press, 2019) on the cosmopolitanism of the lived enlightenment.

32 Livingstone, *Putting*, 23.

33 Henri Lefebvre, *The Production of Space*, trans., Donald Nicholson-Smith (Oxford: Blackwell Publishing, 1991), 183, 191.

read"; "'reading' follows production in all cases except those in which space is thus produced especially in order to be read."[34] It may then be argued that interpretation always follows the inherently social perceptions within space – that is, a geography of knowledge.

5   Geographies of Knowledge: Place vs. Space and Places of Accumulation and Control

Geography and the geographies of knowledge have noted a distinction between space and place. For the "phenomenological" turn in geographies of knowledge, scholars highlighted "place" as a lived experience of "being there" as opposed to the abstract mathematical (Cartesian) space of theoretical science, which we associate with "extension." Cartesians (and some Galileans) imagined the mathematical substrate of measurable extension. For Newtonians "space" is that into which matter was thrown – matter is placed. *Pace*, for the phenomenological movement, *place* mattered.[35] The abstraction of "space" came as a *consequence* of dwelling in a particular mathematical/empirical space, a place. For Anthony Giddens, "[p]lace is best conceptualised by the idea of 'locale' which refers to the physical settings of social activity as situated geographically."[36] For Charles Withers, "place" is used as a framing device in three related ways: first as a collecting site, second as an ordering term (a process in which the world and its products are mapped, named and classified), and third as places of display.[37] Yet even geographers of knowledge often mix up both terms – when they say "space", they mean "place".[38]

But place also has space. Science (knowledge) moves about and goes places. And it is that tracking of movement that first inspired historians of science to "follow science around" as it moved from place to place – first in the diffusionist models, but then in the more circular roundabouts, trailing along with

34   Lefebvre, *Production of Space*, 143.

35   The classic statement is in Yi-Fu Tuan, *Space and Place: The Perspective of Experience* (Minneapolis: University of Minnesota Press, 2001).

36   Anthony Giddens, *The Consequences of Modernity* (Place: Stanford University Press, 1990), 18.

37   C.W.J. Withers, "Geography, Natural history, and the Eighteenth-Century Enlightenment: Putting the World in Place," *History Workshop Journal* 39 (Spring, 1995): 138.

38   Historian of science Steven Shapin warns against conflating space and place. "Spaciality", he would write, "is a necessary condition for their being such a thing as science (or art or the market economy or football)." Steven Shapin, review of Livingstone, *British Journal for the History of Science* 36 (2003): 89–90.

those go-betweens and accumulations.[39] It is these special accumulations in a place that animated Bruno Latour in his examination of "circles of credit" and "immutable mobilities" of the 1980s.[40] We only need to look at the collectors and activities of the British Museum or Kew Gardens as sites of colonial accumulation to see how science moves around and gets accumulated in a *space*.[41] The site of accumulation mattered.

Here we might usefully consider the manner in which the contemplation of nature morphed into the mastery of it in the early-modern world. This is not simply a technological proposition, nor an instrumental one even after Galileo and his disciples. It does, however, mean that absolute space as its own entity was increasingly replaced by secular space—that is, "freed from politico-religious space…"[42] This new social space converged with nature in the social relations of production with all of its attendant economic consequences. As Lefebvre argues, the space of thought and of action also meant a space of control, production and domination.[43] Spaces for the contemplation of nature were thus spaces for commerce and commodities in an age of industry and empire. Numerous spaces were made simultaneously *into* spaces of science, whether lecture halls in scientific institutions, or on ships decks while circumnavigating the globe, and laboratories in urban centres or country homes, collections of instruments, to name a few.

One of the best of such emergent sites was Joseph Banks' Kew Gardens which was designed to reflect the breadth of the enlightened botanical empire. Banks claimed such a range, as "our King at Kew and the Emperor of China at Jehol solace themselves under the shade of many of the same trees and admire

---

39   Jim Secord got the balls rolling in his "Knowledge in Transit," *Isis* 95 (December, 2004): 654–672. Simon Schaffer et al followed it up with a rehearsal of Harley's "go- betweens", in Schaffer, L. Roberts and J. Delbourgo, eds., *The Brokered World.Go-Betweens and Global Intelligence, 1770–1820* (London: Science History Publications, 2009).

40   Summarized in Bruno Latour, *Science in Action* (Cambridge, Mass.: Harvard University Press, 1988).

41   For the concentration in Britain, see Jim Endersby, *Imperial Nature* (Chicago: University of Chicago Press, 2008), and James Delbourgo, *Collecting the World: Hans Sloane and the Origins of the British Museum* (Cambridge, Mass.: Harvard University Press, 2017.) A useful comparison might lay in the examination of the different attitudes towards "space" (and monopoly, laissez faire exchange, and so much more) between the British and French model of colonial natural collection. See G. McOuat, "Cataloguing Power: delineating 'Competent Naturalists' and the Meaning of Species in the British Museum," *British Journal for the History of Science* 34 (March, 2001): 1–28.

42   Lefebvre, *Production of Space*, 256. Cf. Jacob, *Secular Enlightenment*.

43   Lefebvre, *Production of Space*, 26–27.

the elegance of many of the same flowers in their respective gardens."[44] Of course this was self-serving, if only to reflect Banks' influence at the British Board of Trade in 1778. Remarkably, Banks' ally Johann Rheinhold Forster promoted encounters with others,

> ...therefore let us not despise any of them, though they be our inferiors in regard to many improvements and points of civilisation; none of them is so despicable the he should not, in some point or other, know more than the wisest man of the most polished nation.

It also did not reflect the distinctions of court culture revealed when the Chinese Emperor allegedly sneered at the instruments demonstrated to him when the Macartney expedition arrived in China in 1793.[45] Scientific spaces could be geographically defined, as Charles Withers has shown us.[46] But they could also be local and personal. All spaces are lived and social; just as the body became a space of scientific exploration as much as ships in the Pacific or chemical furnaces in private homes. Most significantly, every instrument employed was an instrument applied not just to objects but simultaneously to our bodies so that, as Michael Polanyi once remarked, "our person expands into new modes of being."[47]

Thus, the tension fundamental to crossing spaces—even when the activity may be construed as inherently local—whether in commerce or in translations. We wish to emphasize the growing relation between new spaces and scientific work. If we were to return to Erasmus of Rotterdam at the intersection of the late medieval and early-modern worlds, we would find a claim that private space is necessary for work—not merely, perhaps, for the soul but for the intimacy required for contemplation.[48] In the period which followed, we emerged from the musings of philosophers to the milieu of knowledge accumulation as much as of the more obvious accretion of commodities and capital associated

---

44   Quoted in Livingstone, *Putting Science in its Place*, 56.
45   Cf. Withers, "Geography, Natural History, and the Eighteenth-Century Enlightenment," 146, quoting Johann Rheinold Forster; Alain Peyrefitte, *The Immobile Empire* (New York: Vintage, 2013), 275; William Jardine Proudfoot, *Biographical Memoir of James Dinwiddie. Astronomer in the British Embassy to China, 1792, '3, '4. Afterwards Professor of Natural Philosophy in the College of Fort William, Bengal* (Liverpool: Edward Howell, 1868), 53.
46   For example, Charles W.J. Withers, *Placing the Enlightenment. Thinking Geographically about the Age of Reason* (Chicago: University of Chicago Press, 2007).
47   Quoted in Livingstone, *Putting Science in its Place*, 80.
48   Bachelard, *Poetics*, 65.

with the growth of towns and empires.[49] This was the transformation from cathedral to urban bourse, from cloister to investment, insurance, and worldwide trade. It was, ultimately, a transit of the spaces of science reflected in the transport of nature in the 18th and 19th centuries. Thus, cartography was its own geographical science, as maps reflected towns, ports and sea voyages of increasing length and danger only to be mastered ultimately by grand triangulations and the skill constructed in the marine chronometer.

This collection of essays explores the erasure of limits of space in Enlightenment science. All, to some extent, point to the ways the physical spaces in which scientific knowledge was encountered also implied ways in which knowledge played a social role in encouraging debate and extending investigation. This is one of the great themes underlying Rob Iliffe's account of William Stukeley's search for spaces of contemplation in the late 17th and early 18th centuries. Stukeley desperately sought out spaces where it was possible to engage in scientific thought, from college to countryside, from Leicestershire to London, where it would be possible to escape the expectations placed on the most engaged natural philosophers but who needed solitude as much as encouragement. Between private and public there was a space which required a careful cultivation and sometimes even avoidance. Similarly, the examination of early chemical issues depended on adopting scientific instruments not only in precisely defined private spaces but also in places which were also transformed by devices and their experimental use. Hence, Margaret Carlyle and Victor Boantza take us into private laboratories and workshops of the translator Marie-Geneviève-Charlotte Thiroux d'Arconville in her 18th century efforts to explain the chemistry of putrefaction. The transition between laboratories and other sites of investigation reveals the way spatial, and even social, limits were overcome.

In many of these essays, instruments figure significantly but also in many different ways. Thus, what may have sometime seemed private experiences also became matters of public debate. As Trevor Levere shows, the instrument collections of Martinus van Marum in Haarlem, the Netherlands, of George Parrot at Tartu in Estonia and of Adam Wilhelm Hauch at Soro in Denmark, revealed the reliance on instruments in the wide range of chemical and electrical researches of the Napoleonic period. These collections were porous and not at all mausoleums; they were museums of energetic activity at the centre of major research debates in the early-modern period. This is also nicely underlined by Alice Marples in her study of Hans Sloane's eclectic, otherwise

---

49   Lefebvre, *Production of Space*, 129.

seemingly undisciplined, collections. These also provided spaces of research notably within 18th century medical debates and in defining the natural history of disease in early-modern Britain.

The uses of instruments were thus also about the uses of spaces. Hence, Jasmine Kilburn-Toppin ventures into legal cases, exploring court records for evidence of how instrument makers managed their own domestic spaces to meet the growing market for devices which often required great skill in their production. Therefore, scientific instruments were not simply devices of consumption, entertainment, and investigation. Their frequent theft also tells us much about how the instrumental market worked and the wealth of knowledge in skilfully-produced apparatus offered off the shelf. This was a scientific dynamic in everyday life. A remarkable demonstration of the function of instruments across spaces is also found in Marie Thebaud-Sorger's venture into the world of public research, in the interplay between open public spaces and the passion for ballooning, especially in early-modern France. She follows numerous adventurers and their experiments through the wide public range of scientific theatre. Here the passion for balloons and the intense debates over pneumatic chemistry intersect in the public sphere. Likewise, Simon Werrett, in his discussion of attempts to deal with the highly-problematic longitude, shows dramatically how the application of rockets in determining place ranged from William Whiston's early 18th century application to navigation and surveying through to J.W.M. Turner's depiction of rockets warning of hidden shoals amidst fogs and storms. From dangers and romance at sea, explosive rockets displayed dramatic results. Rockets were then both devices of investigation and they helped define the space between observatories in Paris and London. Similarly, Stewart sets out to examine science across the space between London's instrument makers, chemical manufacturers, and the world of a scientific demonstrator in Calcutta throughout the late 18th and early 19th centuries. By tracing commerce and communication through a trade in instruments and chemicals, he seeks to trace the transit of knowledge across the space of empire.

These essays together all represent what Lefebvre has called a "code of space."[50] In the growing age of discovery represented here, this collection of essays reflects the reach from the "plane of accumulation....to the plane of production..." It is not simply scientific practice which ensured this. But this reach transformed scientific spaces themselves into spaces for the generation of "basic and absolute truths", even perhaps universal ones, beyond mere "things and objects..."[51] That was the ultimate product of the social spaces of science.

---

50   Lefebvre, *Production of Space*, 269.
51   Lefebvre, *Production of Space,* 218; cf. Lissa Roberts, "Producing (in) Europe and Asia," esp. 860–863.

CHAPTER 2

Escape from Capnopolis: William Stukeley's 'True Academick Life'

Rob Iliffe

The notion that one can think freely and pursue scholarship only by cultivating a *vita contemplativa* has long been a commonplace in Western societies. One Latin term in particular, *otium*, described the protected space and time that was considered essential for academic work. Drawing on classical Roman antecedents, medieval writers balanced *otium*'s association with idleness and moral corruption with the capacity it offered to work virtuously and productively on private political or erudite topics. The *otium* available to Roman statesmen, poets and philosophers who were freed from active, public duties permitted them to reflect on nature, human existence, and the affairs of the world. Although there were many subtle distinctions in their uses of the word, Roman writers generally contrasted an elevated form of *otium* – a "worthy leisure" (*otium cum dignitate*) or a "busy leisure" (*otium negotiosum*) – with licentious or luxuriant idleness (*otium otiosum*). The productive and virtuous use of *otium* was closely bound up with its appropriate physical and geographical settings. While the rural villa was held up as the most significant locale for such activity, many villas existed on the borders of, or actually inside towns and cities. Wherever such buildings were situated, in their libraries and gardens senior politicians, lawyers and other elite citizens could cultivate both a civilised and a scholarly life away from the distractions of the metropolis.[1]

---

1   In general see Jean-Marie André, *L'otium dans la vie morale et intellectuelle à Rome des origines à l'époque augustéenne* (Paris: Presses Universitaires de France, 1966) and B. Vickers, "Leisure and Idleness in the Renaissance: the Ambivalence of Otium (Part 1)," *Renaissance Studies* 4 (1990): 1–37, esp. 6–15, and (Part 2) 107–54. For Roman accounts see Seneca, "On the Private Life" in *Seneca: Moral and Political Essays*, eds. J.M. Cooper and J.F. Procopé (Cambridge: Cambridge University Press, 1995), 167–80 and Cicero, *On Duties*, eds. M.T. Griffith and E.M. Atkins (Cambridge: Cambridge University Press, 1991), 101–2. For the Roman villa see J. Arce, "*Otium et negotium*: the Great Estates, 4th-7th century," in L. Webster and M. Brown, eds, *The Transformation of the Roman World, AD 400–900* (Berkeley and Los Angeles: University of California Press, 1997), 19–32, esp. 22–5.

The advent of Christianity and the emergence of monastic orders forced a revaluation of the benefits of retirement from ordinary society, and contemplative solitude was reconfigured as a condition of attaining knowledge of the divinely created cosmos. Although Church Fathers routinely denounced aspects of *otium* on the grounds of its proximity to sloth (*acedia*), monastic writers argued that *otium* was acceptable when it gave occasion to serve God via the *vita contemplativa*. In *The City of God* and other writings, Augustine argued that a monastic life devoted to the knowledge of God required a holy leisure (*otium sanctum*). This view became widely accepted in monastic communities and as a result the first major schools of learning in Europe were attached to religious institutions that were walled off from the surrounding urban environment. While the latter was a place of trade (*negotium*), sociability, argument, and deceit, study in either the monastery or the countryside was understood to be requisite for gaining knowledge of divine truths. These places were not merely symbolic locations, but really existing spaces in which members of religious orders engaged in both religious and secular studies.[2]

By the twelfth century, independent masters were already teaching students outside the monastic walls. In Paris, "independent" masters armed with a *licentia docendi* plied their trade by drawing income from teaching the liberal arts in their own homes, or in places of business. Regular masters and senior members of the ecclesiastical hierarchy condemned them as word-sellers (*venditores verborum*) who peddled techniques of *disputatio* and inflated the authority of Aristotle and other pagan writers. Faced with such professional and intellectual threats, the Abbot Bernard de Clairvaux told students in 1140 to flee Babylon and "fly to the cities of refuge where you can repent for the past, live in grace in the present, and confidently await the future. You will find much more in forests than in books. The woods and the rocks will teach you much more than any master." Bernard could rail against such practices as much he liked, but disputation and Aristotelian texts would become the mainstay of Arts faculties in the first universities.[3]

Although the university remained the most significant institution for teaching and learning, in the Renaissance growing numbers of scholars associated themselves with the worlds of aristocratic or church patronage. In the fifteenth

---

2 B. Vickers, "Leisure and Idleness in the Renaissance: the Ambivalence of Otium (part 2)," *Renaissance Studies* 4 (1990): 107–54; Georges Duby, "Solitude: the Eleventh to Thirteenth Centuries," in Duby and Philippe Ariès eds, *A History of Private Life*, 5 vols (Cambridge, Mass.: Belknap Press, 1988), vol. 2, 509–33.

3 J. Le Goff, *Intellectuals in the Middle Ages* (Oxford: Blackwell, 1993), 20–4 (and p. 21 for the citation from Bernard), 61–7 and 161–6.

century many humanists looked back to the classical ideal of *otium* by attaching themselves to villas such as that of the Medici at Careggi. Providing a novel setting for scholarly solitude, the library became a central resource for academic *otium*. While representations of the thirteenth-century master depicted him surrounded by students and benches, the fifteenth-century humanist was portrayed (often in the form of St. Jerome) as perfectly alone in a private study full of books.[4]

The most influential rendering of *otium* for Renaissance readers was given by Francesco Petrarch, who emphasised both classical and Christian conceptions of the retired life in a number of works. His *De vita solitaria* (c. 1345–66) was addressed to Philippe de Cabussoles (Filippo di Cabassole, Bishop of Cavaillon) during a period of fertile leisure spent at his retreat outside Avignon in the Vaucluse. For Petrarch, who explicitly followed the advice of Bernard, the proper setting for *otium* was in a countryside, near to woods, fields and streams. He described *otium* as a state characterized by the absence of worldly cares or concerns, and to be attained through the cultivation of self-restraint, study, reflection and writing. This was a condition of mitigated solitude, during which one needed to have access to a library and to like-minded friends, lest one turn in on one's self and fall prey to melancholy – or worse. This type of solitude created the conditions for serious intellectual labour, leading to a revitalized body, virtuous behaviour and knowledge of God. In other writings, such as his *De oto religioso* (1347–60), Petrarch praised the monastic contemplative life as a crucial source of Christian rest, but he frequently drew from classical accounts of the benefits of *otium*. One of his favourite texts was the famous remark attributed by Cato (and cited in Cicero's *De Officiis*, 3:1) to the third-century BCE General, Scipio Africanus the elder, to the effect that he was never less idle than when at leisure, and never less lonely than when alone.[5]

---

4 Le Goff, *Intellectuals*, 166. It should be emphasised that the depictions represented ideals, and also that the 'solitary' scholar was nevertheless in conversation with great minds of the past – see W. Liebenwein, *Studiolo: Die Enstehung eines Raumtyps und seine Entwicklung bis um 1600* (Berlin: Gebr. Mann, 1977); D. Thornton, *The Scholar in his Study: Ownership and Experience in Renaissance Italy* (New Haven and London: Yale University Press, 1997) and A. Grafton, *Commerce with the Classics. Ancient Books and Renaissance Readers* (Ann Arbor: University of Michigan Press, 1997).

5 See E. Kantorowicz, "The Return of Learned Reclusiveness in the Middle Ages," in Kantorowicz, ed., *Selected Studies* (Locust Valley, NY: J. J. Augustin, 1965), 339–51; G. Constable, "Petrarch and Monasticism," in Aldo S. Bernardo, ed., Francesco Petrarca: Citizen of the World. Proceedings of the World Petrarch Congress, Washington,D.C., April 6–13 1974 (Padua and Albany: State University of New York Press, 1980), 53–99; D. Yocum, "*De oto religioso*: Petrarch and the Laicization of Western Monastic Asceticism," *Religion and the Arts* 11 (2007):

For most scholars however, fantasies of academic retirement clashed with reality. In his study of the scholarly life of early modern German academics, Gadi Algazi has shown how in the late fifteenth century, an increasing number of philosophers and theologians (and not merely doctors and lawyers) married and lived in town houses. This required a radical reworking – though not a complete rejection – of the traditional ideals of celibacy and solitary contemplation as the basis of the scholarly identity. In this new environment, finding the space and time necessary for intellectual labour was a fraught business, and one solution was to build a private "study". In Erasmus's 1528 dialogue *The Ciceronian*, the stoic character recommended that married scholars should work at night in a special room protected by impenetrably thick walls, so that they might be undisturbed by blacksmiths, snoring or women. Stories circulated in the mid-sixteenth century about Philip Melanchthon. The great theologian had allegedly locked himself away in his study to prevent interference from his wife, who had knocked continually on his door imploring him to open it. According to similar accounts, his children were apparently welcomed into his fortress of solitude.[6]

By contrast with these paeans to solitude, seventeenth-century empiricist and experimentalist philosophies celebrated a collaborative *vita activa*. Francis Bacon and his followers denounced as unhealthy and barren philosophical methods that emphasised solitary paths through seas of thought, and they assailed the textual focus of quasi-monastic college institutions. Yet being overburdened with worldly affairs invariably left insufficient time to think or to complete philosophical projects. Robert Hooke, the most prominent experimentalist at the Royal Society, Gresham Professor of Geometry and Surveyor for the City of London (after the Great Fire of 1666), frequently complained that he had insufficient time to complete his own works, and thus secure his intellectual property. His excuse of "busyess" was ridiculed by a number of his contemporaries at the Royal Society, although all of them had more access to *otium* than he did. As Steven Shapin has pointed out, all natural philosophers agreed that they needed to execute a temporary withdrawal from society in order to acquire the *otium* necessary for fruitful scholarly work. Indeed, in 1622 Bacon himself had told Elizabeth, Queen of Bohemia, that his chief desire was "to haue leasure without loytering, and not to become an Abby-lubber as the old proverb was, but to yeeld some frute of my priuate life." For Robert Boyle,

---

454–79; J. Bordanella, "Petrarch's Rereading of *Otium* in *De Vita Solitaria*," *Comparative Literature* 60 (2008): 14–28, esp. 14, 16, 19–20 and 23–4.

6   G. Algazi, "Scholars in Households: Refiguring the Learned Habitus, 1480–1550," *Science in Context* 16 (2003): 9–35, at 11–14, 17–20, 25–8 and 30–2.

setting aside afternoons where he would be undisturbed by "unwelcome visits" was the only way both to perform experiments thoroughly and to write up his results.[7]

There were a variety of ways to escape the constraints on time provided by the city. Universities were never as stifling or reactionary as their critics claimed, although many of the heroes of the Scientific Revolution were more able to promote novelties and disrupt disciplinary hierarchies outside them. Yet the collegiate environment continued to serve both symbolically and in reality as a quasi-monastic retreat from the evils of the city. During his career at Cambridge, Newton often contrasted the freedom he got from being in a college with the allegedly burdensome demands placed on him by his obligations to the Royal Society of London and more generally, by the Republic of Letters itself. When it suited him, he also claimed that being in the countryside – either in Cambridge or Woolsthorpe – left him ignorant of what was taking place in London. John Locke cast his own disengagement from London differently, citing the hideous smoke of the "Town" (which brought on attacks of asthma) as much as the benefits of study in a country house as the key reasons for living outside the metropolis. Following his return from his Dutch exile in early 1689 he no longer had an inclination to return to his *alma mater* (Christ Church) at Oxford, and instead lodged first at Parsons Green, at the earl of Monmouth's Villa Carey, and then at Oates near Harlow, at the home of Francis and Damaris Cudworth Masham.[8]

In all these examples a complex set of relationships existed between the values associated with urban, monastic and rural locales. In the rest of this paper

---

7   S. Shapin, "'The House of Experiment in Seventeenth-Century England," *Isis* 79 (1988): 373–404, esp. 386–7 (for Boyle's escape attempts); idem, "'The Mind is its Own Place': Science and Solitude in Seventeenth-Century England," *Science in Context* 4 (1991): 191–218, esp. 197–8 and 210–11. For Hooke see e.g. Newton to Halley, 20 June 1686, in H.W. Turnbull et al., *The Correspondence of Isaac Newton*, 7 vols (Cambridge, 1959–77), vol. 2: 438; Bacon to Elizabeth Stuart, 30 April 1622, in N. Akkerman, ed., *The Correspondence of Elizabeth Stuart, Queen of Bohemia*, 3 vols (Oxford: Oxford University Press, 2015), vol. 1: 357.

8   See for example, Newton to Hooke, 28 November 1679 in H. W. Turnbull ed., *The Correspondence of Isaac Newton* (Cambridge and New York: Cambridge University Press, 1960), vol. 2: 300; the earl of Monmouth to Locke, 10 October and 19 November 1692 in E.S. de Beer, ed., *The Correspondence of John Locke*, 8 vols (Oxford, 1976–89), vol. 4: 527–8 and 583. More broadly see Shapin, "'Mind is its Own Place'," esp. 198–9 and 203; B. Vickers, "Public and Private Life in Seventeenth-Century England: the Mackenzie-Evelyn debate," in Vickers ed., *Arbeit, Musse, Meditation: Betrachtungen zur Vita Activa-Vita Contemplativa* (Zurich: Verlag der Fachverine, 1985), 257–78. For the Renaissance court culture as the quintessential site for innovative thinking see R.S. Westman, "The Astronomer's Role in the Sixteenth Century," *History of Science* 18 (1980): 105–47.

I look at the efforts of the great eighteenth-century "Antiquary" William Stukeley to locate the appropriate setting in which to obtain the *otium* necessary to conduct his scholarly activities. Over half a century he left frank accounts of his successes and failures at locating such an Elysium. Stukeley began by expressing a distinct and classically Augustan preference for a rural setting, with a villa, study and accompanying garden specifically designed to create the conditions for productive academic work. However, London, where he would live for various periods of his life, was a wonderful source of conversation, and the ever-sociable Stukeley joined a variety of clubs and gatherings, founding a number of societies when the occasion arose. Nevertheless, he understood urban life to be a hotbed of libertinism and irreligion, and it was always physically and mentally debilitating. While coffeehouse converse was exhilarating, Stukeley's need for an income meant that he had neither time nor space for the life of the mind. Against this, the countryside, to which he periodically retreated for tours with friends during his London life, was healthier for the constitution, and – at least aspirationally – it offered him the only realistic prospect of pursuing his studies. He frequently observed that it was only in such a locale that he could truly be free to pursue a pleasurable but useful life of "curiosity".

Stukeley moved wholesale to the countryside at two key stages of his life. As it turned out, the rural environment by itself could not meet his expectations, and fearing that its solitude left him exposed to a living death, he built into it different sorts of sociability, based to some extent on what he had experienced in London, in the form of literary societies or masonic lodges. On the other hand, when he returned to live in the Capital, as he did in 1717 and 1748, he strove to generate the requisite space and time to lead what he called a true life, freed from the debilitating demands of society. Only in his seventh decade, when he had retired from his career as a physician, had become independently wealthy, and had moved to the dreaded metropolis, was he able to live a life that struck the right balance between his otial and sociable aspirations.

1   An Inclination to Retirement

Stukeley was born in 1687 at Holbeach in Lincolnshire, the son of a local lawyer and a supportive mother. His house had 10 acres of land, and his father stocked it with trees and an orchard, as well as with a stable and various fishponds. In one of a series of recollections about his youth, Stukeley recorded that he was fascinated by antiquities, devoted to exploring local surroundings and deeply interested in local wildlife. He remembered that his early pursuits were not the merely playful activity of country boys, but a serious examination of nature.

Even as a youth he was conscious that the retreat to the countryside provided an ideal locale for reading books and for thinking seriously about the natural world. At school, he later recalled,

> I was always possessed with a mighty inclination of retiring into the Woods & little shady places in the Parish & round about, so that on holy-days I generally passed a good deal of time there, & whilst the other boys were busy in hunting for birds nests, I busyd myself in reading some book I carried in my pocket, or contemplating the shrubs & plants.[9]

Stukeley also gleaned practical information from a number of local people, often friends or clients of his father, who were willing to pass on words of wisdom or some of the secrets of their trade.[10]

At the age of 13 Stukeley's father began to include his son in some of his business dealings, with the intention of having William entered into the Inns of Court to study for the law. From the summer of 1701, Stukeley accompanied his father to London at regular intervals, but the boy was too concerned with his schoolbooks and "whatever little time I had to spare I generally spent in viewing the buildings, monument, & frequenting Booksellers shops." Instead of following his father's wishes and "hearing the tryals" at Westminster, he spent a great deal of money on instruments and bought many books on science and medicine, "which at all leisure hours I was continually poring upon, & drawing schemes from." He spoke to the men who were building St. Paul's Cathedral, and long before Health and Safety regulations were imagined, climbed the scaffolding that then reached as high as the bottom of the great dome. Away from the Town, he gave vent to his propensity for studying astronomy, medicine and natural philosophy, and for examining old coins and objects, and he toured the countryside for medicinal plants (a practice known as 'simpling') with fellow students such as Stephen Hales. Realising that a legal career for his son was a lost cause, and believing that William might prepare for the cloth, his father allowed him to go to Corpus Christi College at Cambridge, where he arrived in November 1703.[11]

---

9   W.C. Lukis, ed., *The Family Memoirs of the Rev. William Stukeley, M.D. and the Antiquarian and Other Correspondence of William Stukeley, Roger & Samuel Gale, etc.,* 3 vols (Edinburgh: Surtees Society, 1882–7), vol. 1: 13–14.

10 Lukis, *Memoirs,* 1: 13–14. See more generally S. Piggott, *William Stukeley. An Eighteenth-Century Antiquary* (London: Thames and Hudson, 1985) and D. Haycock, *William Stukeley: Science, Religion and Archaeology in Eighteenth-Century England,* (Woodbridge, Suffolk, 2002), esp. 30–43.

11 Lukis, *Memoirs,* 1: 17 and 50.

At Cambridge Stukeley was in his element, spending the whole of his first year at the university and immersing himself in books, lectures, natural philosophy, dissections, and simpling. He walked or rode around the surrounding countryside, finding in his first two years there that he had a particular penchant for the study of medicine. Even within the confines of the city, there were places of retreat where he could explore or read. Later, in 1720, he recalled that

> busy in not ignoble leisure, I had fully determined my thoughts to the study of Physic, & felicitated myself upon it, perceiving the noisy bar would never have been for my purpose, or consentaneous to my invincible modesty & want of assurance … I took great delight in going into St Johns Gardens & studying there [and] I judgd that I could better argue upon paper, if occasion was, than viva voce, and that my temper would never suit with that tumultuous manner of reasoning & rugged kind of study.[12]

When Stukeley did return to Holbeach his father surrounded him with learned people and other good company, fearing he might spoil his chances of becoming a "considerable Man" by contracting bad country habits such as drinking to excess. He spent many hours conversing with his friend Abraham Pimlow, telling him in April 1705 that it was by good providence that he had been taken "from that troublesome, & laborious, though gainful state" of practising law. Instead, he had been brought "into this renowned theatre of learning & wisdom," adding that he hoped that on his return to Holbeach, since they shared the same love for retirement and study, "we shall in some wise imitate an Academick life in the country."[13]

At the university, Stukeley was able to access the *otium* that was required for a life of study. However, fate was soon to intervene in a disastrous manner, and he was inexorably thrust into the urban world that he so detested. In February 1706 his father died in his London chambers from complications, as Stukeley saw it, of his chronic tendency to gout. Ironically, Stukeley heard the terrible news while simpling in fields near Newnham, enjoying "the most thoughtless and serene of any part of my life." His paternal uncle passed away from grief a few weeks later, and in the summer his youngest brother died of smallpox. Dealing with the terrible issues that followed his father's demise

---

12   Lukis, *Memoirs*, 1: 24.

13   Lukis, *Memoirs*, 1: 27; Stukeley to Pimlow, 17 April 1705, in Lukis, *Memoirs*, 1: 142–3. Original is Bod. Eng. Misc. c. 533

would dramatically alter Stukeley's plans, but initially he was able to continue his academic existence. After a bout of smallpox he spent the summer of 1706 touring the countryside, dissecting animals, simpling, and studying antiquities with doctors from neighbouring towns. Expanding his interest in the distant past, and giving vent to his great talent as an artist, he made sketches of King's College chapel and the ruins of Barnwell Abbey.[14]

Stukeley attended lectures as usual at Cambridge early in 1707, but tragedy struck again that summer, as first his mother died, and then within a month another brother succumbed, this time to dysentery. Aside from the trauma of his losses, Stukeley now had to act fast to pay off the substantial debts incurred by his father, but as he put it, in the bulk of these he "was cheated off thro' my own unskilfulness in such affairs & my avocations to follow my studys which I was resolved not to neglect at all events, tho' expensive." As this reminiscence suggests, none of the personal disasters that befell him put a complete stop to his simpling or to his burgeoning antiquarian pursuits, which he pursued throughout 1708 with a growing band of fellow enthusiasts. Over the winter of 1708–9 he prepared for the examination for his Bachelor of Physic degree, which he successfully passed in January 1709. As soon as he could, he escaped to the country and made good use of his time to continue his interests. In April 1709, for example, he spent time with a Mr Lucas and his sisters at Holywell in Northamptonshire, living in "a kind of Monastic communication" with the women and voyaging around local ruins "like Errant Vertuosos" with one of the sisters.[15]

2   The Spoils of the Dead

In August 1709 Stukeley set out for London to complete his medical studies and to prepare for practice by spending time in hospitals learning about the cause, progress and treatment of various diseases. En route he was struck down by gout, which he believed he had inherited from his father. He would later associate the occurrence of the disease with his London lifestyle, and he would always frame his retirement from the English Babylon as an escape from metropolitan ill-health. For seven months he worked at St. Thomas's Hospital under the guidance of the great physician Richard Mead, and attributed much of his subsequent travails with gout to the copious amounts of French wine they quaffed. He soon grew "heartily tird" of London and made a firm resolution

---

14   Lukis, *Memoirs*, 1: 26 and 29–32.
15   Lukis, *Memoirs*, 1: 34–5, 38–9 and 41–5.

to retire to the country. In February 1710 he told Pimlow that he was "so weary of this noisy, stinking Town, that I think long to be in the country" and soon after this he moved to Boston in his home county. Many of his relatives lived in the town, and he began to build up a practice that would help pay off his remaining debts. Here again he was impressively energetic as a naturalist and displayed the first signs of the creative aspect of his sociability by launching a "'botanic club". From 1711 he spent many weeks each year simpling with local apothecaries, and joined the recently founded Gentleman's Society of Spalding. His friendships with its founder, Maurice Johnson, and with the two Gale brothers Roger and Samuel, all of whom he had met in London, would be the most important intellectual relationships of his life.[16]

The lure of the city exercised a powerful counterbalance to the delights of rural retreat. Stukeley returned to London at Ormond Street in May 1717, later recalling that his rationale for returning to the metropolis had been to "lead a life of study & curiosity." In reality, it was the result both of necessity and a basic calculation of means and ends. A successful medical career promised to bring him the sort of money that could set him up as a gentleman, but more importantly it would guarantee him the leisure time to pursue his travels and studies at certain times of the year. In June 1717 he told Maurice Johnson that his income in Lincolnshire had been insufficient, "& instead of dirty roads & dull Company I need not tell you what we meet with in London." He added that he enjoyed the perfect 'Rus in Urbe', "& in my Study backwards I have a fine view to Hampstead, & the rural scene of haymakers, Mugitusque boum." Not for the first time Stukeley would indicate his textual and existential indebtedness to the peaceful and "natural" agricultural life portrayed by Virgil in his *Georgics*.[17]

With the help of Mead, Stukeley started a popular practice in Great Ormond Street, though his income was never enough for him to support an academic life. It was certainly far from the lifestyles of those physicians like Mead and

---

16   Ibid., 1: 46–7, 51–2; Stukeley to Pimlow, 4 Feb. 1709/10; ibid., 1: 149; Haycock, *Stukeley*, 44–45. For his ill-health due to gout, and his subsequent expertise in the disease, see K.J. Fraser, "William Stukeley and the Gout," *Medical History* 36 (1992): 160–86. For Stukeley and Johnson see *The Correspondence of William Stukeley and Maurice Johnson 1714–54*, eds. D. Honeybone and M. Honeybone (Lincoln: Lincoln Record Society, 2014), and for eighteenth century correspondence in the Spalding Society see *The Correspondence of the Spalding Gentlemen's Society 1710–61*, eds. D. Honeybone and M. Honeybone, (Lincoln: Lincoln Record Society, 2010).

17   Lukis, 1: 122; Stukeley to Johnson, 13 June 1717, in Honeybone and Honebone, *Correspondence of Stukeley and Johnson*, 24. The reference to the lowing of cattle is from *Georgics* Bk 2 line 470.

Sloane whose ostentatious displays of wealth he later professed to despise. London did allow Stukeley to meet the most talented and prestigious scholars in Britain, and he participated fully in its social and intellectual milieus. Living in London allowed him to give free rein to his tremendous organising energy, and he created a number of clubs for virtuosi of different kinds, including short-lived examples in Avemary Lane and Orange Street. He was a prime mover in instituting the Society of Antiquaries with Roger and Samuel Gale on 1 January 1718, and three months later he was made a fellow of the Royal Society. For this he had the support of Mead and Isaac Newton, the latter another Lincolnshire lad and before long, a significant social acquaintance. With Mead as patron, his academic career also took off. He took his doctoral degree in medicine in July 1719, and was admitted into the Royal College of Physicians in September 1720, giving the Gulstonian Lecture on the subject of the spleen in 1722. At the same time, he gained a reputation for expertise in antiquarian pursuits and natural history. It was on the basis of publishing a paper on a Roman monument in Scotland that he was invited to become a freemason, an offer he accepted in January 1721, and at the end of the year he was made Master of a new lodge created at the Fountain Tavern in the Strand. Soon afterwards, in 1722, he helped found the Society of Roman Knights, a group open to female members that was dedicated to appreciating Roman culture and to restoring the glory of Rome in Augustan Britain.[18]

At the gatherings of these societies Stukeley regularly met and counted himself as a friend of many powerful grandees. Later he noted that he had known many of the greatest men in London in the nine years he was there, "& by having recourse to their librarys I arriv'd to a considerable degree of knowledg & equal reputation." For nearly a decade Stukeley threw himself into the metropolitan world, periodically escaping into the book-filled recesses in the homes of great men. He consolidated his position in the many London societies of which he was a member, trying to avoid coffeehouses where infidels and freethinkers spread their materialist poison. Gout dominated his life in London, and debilitated him each year around March, apart from 'remissions' in 1719/20 and 1720/1. It struck him heavily in 1722 and then again in the same month in the following two years. It surfaced more regularly in the mid-20s, in December

---

18   Lukis, *Memoirs*, 1: 51, 77, 122; Haycock, *Stukeley*, 116–20. Stukeley would develop an ambivalent opinion of Mead, praising his friendship and talent, but lamenting his pursuit of lucre and descent into irreligion; see ibid., 1: 111–12. For the location and proceedings of the Society of Antiquaries, see Roger Gale to Sir John Clerk, 26 April 1726; Lukis, *Memoirs*, 1: 181–2. On the vogue for local societies in this period, see P. Clark, *British Clubs and Societies 1580–1800. The Origins of an Associational World* (Oxford: Oxford University Press, 2000).

1724 and 1725, and in May 1725 and 1726, though by now he had learned that bathing his gouty foot in cold water, sobriety, and drinking lots of water got rid of the symptoms within a couple of days. In his horoscope he noted that gout was caused by lack of country air and exercise and, to address this, in the spring and summer he would leave the Capital and ride around parts of England on horseback. Often, he toured the countryside alone but sometimes travelled in the company of Roger Gale, both losing themselves in past worlds and cultures about which he and Gale acquired increasing amounts of expertise.[19]

Stukeley's first book-length work on antiquities, the *Itinerarium Curiosum* of 1724, cemented his reputation as an antiquarian but its lavish plates made it prohibitively expensive and it sold badly. Although he had harboured a desire to travel on the Grand Tour, he now reconciled himself to the likelihood that this would never happen. Instead, he recast the British landscapes he explored and recounted in the *Itinerarium* as superior to, and some cases older than anything that could be found in Continental Europe. He had read a number of books on British antiquities and sites of interest, but in the *Itinerarium* he emphasised that he had personally travelled to the places he described. He gave pithy but laudatory accounts of old roads, fountains, burial grounds, Roman mosaics and the ruined remains of pre-Reformation religious buildings, pausing when he could to comment on the beauty of the landscapes and the fertility of their soils. In the second half of his book he added a large number of pictures of the places he had visited, arguing that "Proper engravings" provided a wonderfully vivid and edifying idea to the mind. Indeed, Stukeley's lifelong penchant for drawing different sorts of landscapes, notably aspects of his own properties, constitutes abiding testimony of his attachment to the power of place. At the outset of the work, he praised the opportunities for study and conversation in London, but he lamented the fact that confinement within it for too long was insupportable, giving rise only to an "artificial" life. Instead, he needed "that real life, that tranquillity of mind, only to be met with in proper solitude." In such a state one could store the mind with "valuable treasures of the knowledge of divine and natural things," which was fulfilled best by studying antiquities. In this way, the book was supposed to be a guide for the "Curious", who could obtain pleasure as well as edification by travelling throughout Britain. There were "rarities of domestic growth," Stukeley noted, and it was possible "to make a classic journey this side of Dover."[20]

---

19   Lukis, *Memoir*, 1: 51–4, 69–75, 92. For his self-treatment, see in particular Lukis, 1: 69–73 and Fraser, "Stukeley and the Gout," 165–8.

20   Stukeley, *Itinerarium Curiosum. Or, an Account of the Antiquitys and Remarkable Curiositys in Nature or Art ...*, (London: For the author, 1724), 1–3 and Haycock, *Stukeley*, 109–116. For

P. 40   TAB. XXI

Stukeley del.

An inward View of STONEHENGE. AA. *the altar*
or Side view of the cell.

FIGURE 2.1   View of the 'altar' or adytum of stonehenge, plate XXI from Stukeley, *Stonehenge.*
A temple restored to the British Druids, London, 1740. Courtesy of the Bodlein
library, public domain.

From early on Stukeley was impressed by the great age of the Neolithic
structures he encountered. Having procured a copy of John Aubrey's manuscript account of Stonehenge, he visited Stonehenge and Avebury with Roger
Gale in the middle of May 1719 and remained obsessed with these sites for the
rest of his career (figure 2.1). In a typical observation, composed as part of an
early 1720s essay on the ancestry and function of Stonehenge, he noted that it
was set in a stunning landscape, and was at least as magnificent as anything in
Egypt. He disagreed with writers such as Inigo Jones and John Webb, who had
claimed that it was a Roman construction, and he noted that it was "a glorious
piece of Antiquity" of immense age. Early on in his researches he came to the
conclusion that these and similar sites were fashioned according to the units
and standards of measurement used in the Egyptian pyramids and the Temple
of Solomon. As part of the ancient patriarchal religion, it shared the round
form common to the oldest temples in the world, and the *sanctum sanctorum*
or *adytum* was the central part of a grand Cathedral. The thrust of Stukeley's
argument was not merely to prove that Stonehenge could not have been built

interest in the ancient British past see R. Sweet, *Antiquaries. The Discovery of the Past in*
Eighteenth-Century Britain (London and New York: Hambledon, 2004) and S. Smiles, *The*
Image of Antiquity: Ancient Britain and the Romantic Imagination (New Haven: Yale Uni-
versity Press, 1994).

by the Romans, and certainly not by the Danes and Saxons, but to demonstrate that it was designed and constructed by indigenous British Celts, who drew on Egyptian aesthetics and construction techniques. The architects of the structure were men "who livd in colleges & communitys separate from the vulgar", and who created such edifices in order to increase the veneration of the ordinary people towards them.[21]

3   The Perils of Fumopolis

Although London social and intellectual life was seductive, Stukeley continued to crave the opportunities provided by the country life that was accessible for only limited periods during his expeditions. The situation came to a head at the end of 1725, while touring Newcastle with Roger Gale. Contemplating the Milky Way one night, Stukeley recalled, "an irresistible impulse seiz'd my mind to leave the Town" and he decided to retire to Grantham. By the time he left London in the summer of 1726 he was a major figure in the Royal Society, the Royal College of Physicians and the Society of Antiquaries, and as he put it, "in the highest favour with all the great men for quality, learning, or power." Nevertheless, at this very moment, he decided to leave for the country, "a resolution thought of by many, executed by very few," and carried out "to the wonder and regret of all his acquaintance." Later in his career, Stukeley cited various reasons for leaving the metropolis, never settling on one single justification for his move. These accounts variously cited his deteriorating mental and physical health, his desire to move closer to his brother, his continued predilection for the pleasures and comforts of rural life, his inability to generate the time and space necessary to pursue his antiquarian studies, his desire to study the "tenets, & mysterys of those old philosophical priests of the patriarchal religion" (i.e. the druids), and the operation of Providence. Contemporary letters and other writings reveal that there were many different motivations underpinning the move, and there was no overriding reason to leave beyond a general sense that the countryside was a better setting for an authentic scholarly life.[22]

---

21   A. Burl and N. Mortimer, eds., *Stukeley's 'Stonehenge'. An Unpublished Manuscript, 1721–24* (New Haven and London: Yale University Press, 2005), 27, 38–9, 52–62, 70–3, 84, 118–22 and 127–8; see also Haycock, *Stukeley*, 119, 123–5 and 128–32.

22   Lukis, *Memoirs*, 1: 52–3, 105–8 and 123, and Stukeley to Sloane? December 1726; Lukis, *Memoirs*, 2: 259. Stukeley later considered that it had been a mistake to move to Grantham in order to be closer to his brother, for "one commonly finds less friendship among relations than others"; see ibid., 1: 123.

Moving to the countryside was not in itself enough to provide the *otium* that Stukeley needed, and he did not yet have sufficient funds to support the existence to which he aspired. He told Roger Gale in April 1729 that he agreed with Seneca (presumably *Epistles* 72–3) that business was a "great devourer" of time, and was only undertaken by those who had no opportunity to spend their time better. Stukeley knew that a satisfactory income was a necessary means to this end, but he was well aware of how destructive the excessive attachment to money could be. He had invested in South Sea stock in the spring of 1720 and in his diary he recorded the madness of the people that resulted from the bursting of the Bubble, an episode that typically allowed him to reflect on the baneful effects of metropolitan existence. Trading in Exchange Alley, he wrote,

> increased my distast to business. Nor could I bear the loss of time, which it necessarily brings upon us, even when we have nothing to do. For the same circle must be observed every day of one's life, like a horse in a mill, and one's head must be constantly filled with the empty nonsense of coffee house chit-chat, and public company ... I saw my brethren of the college inriching themselves with the spoils of the living. I coveted only those of the dead. I despised their gaudy life and sumptuous entertainments.[23]

Of his London period he later noted that he could have become rich, but his early hardships had made him "indifferent to a lucrative profession." He preferred wisdom to riches, and concluded that "My love of learning engag'd me into the sweets of a contemplative life."[24]

The contrast Stukeley offered between the ills of the city and the benefits of the countryside, or more particularly, of the garden, was of course a common trope.[25] London in particular was routinely denounced as a haven of avarice,

---

23 Lukis, *Memoirs*, 3: 461.

24 Lukis, *Memoirs*, 1: 62, 64, 79 and 122; Stukeley to Roger Gale, 22 April 1729, ibid., 2: 264.

25 See for example, M. Mack, *The Garden and the City. Retirement and Politics in the Later Poetry of Pope, 1731–1743*, (London: Oxford University Press, 1969). For negative depictions of the environment of early modern London see L. Manley, "From Matron to Monster: London and the Languages of Description" in *Literature and Culture in Early Modern London* (Cambridge: Cambridge University Press, 1995), 125–167; M. Jenner, "The Politics of London Air: John Evelyn's 'Fumifugium' and the Restoration," *Historical Journal* 38 (1995): 535–51; L. Williams, "'To Recreate and Refresh Their Dulled Spirites in the Sweet and Wholesome Ayre': Green Space and the Growth of the City," in J.F. Merritt, ed., *Imagining Early Modern London. Perceptions and Portrayals of the City from Stow to Strype, 1597–1720* (Cambridge and New York: Cambridge University Press, 2001), 185–216 and W. Cavert, *The Smoke of London. Energy and Environment in the early Modern City* (Cambridge: Cambridge University Press, 2016), esp. 213–31.

luxury and envy -- as Stukeley put it, it was a "never ceasing round of pleasure, show & entertainment." The health benefits of the countryside, which stood in stark contrast to the deleterious effects of the city, were central to his accounts of his move. All the efforts to gratify the artificial impulses of the city were debilitating; there were many diseases of the body and anxieties of the mind, and the urban dweller was permanently dissatisfied. On the other hand, the countryside offered real contentment, since true happiness consisted in wanting little, rather than in possessing much. Excessive consumption was a major reason for Stukeley's departure, since it gave rise to recurrent fits of gout, which he noted at one point, "I thought to check by country air & exercise." Initially he claimed that he was satisfied with this therapy, but when he moved to Stamford in 1730 he came upon what was soon a widely used oil-based cure for the disease discovered by one of his parishioners, Dr. John Rogers. It was providence, Stukeley argued, that had prompted him to come to Stamford just at the time that Rogers invented the cure, and which had then allowed Stukeley to publish it for the benefit of all. London had also exacted a toll on his mental powers, and in another of his reminiscences he noted that city life had driven him close to mental despair. He added that immediately after he had left the Capital he had bought a house in Grantham, "finding no purpose in life can be answered by my stay in London; but study, & too close attachment to that, had like to have thrown me into a hypochondriacism."[26]

4   Grantham: Useful Pleasure and the Sylva Academi

Stukeley raved to his friends about the benefits of his Grantham existence, and immediately set about transforming his property into an abode fit for a true philosopher. In February 1727 he told Samuel Gale that he regretted having previously been sceptical of Gale's great endorsement of the "innocent pleasures" of the country life. Now he had become a passionate devotee of rural nature, which provided a solid contrast to his "gilded prospect of imaginary enjoyment" in the metropolis. Previously, his reasoning had been based

> upon the conceit that there can be no conversation worthy of a man of sence but at London. There alone we have the liberty to expatiate & shine in the several qualifications genius or study & application make us masters of. There only we can meet with souls tuned up to our own

---

26   Lukis, *Memoirs*, 1: 53–4, 77–8, 106–8 and 2: 259–60. For Rogers see Fraser, "Stukeley and the Gout," 168–76.

pitch, & indulge the pleasure of taking & giving instruction, of improving & enlarging our ideas, & revelling in mental rapture; & 'tis not to be
denied but that in a great measure such is the case; & no one was more
sensible of it than myself, or rioted more in the luxury of conversation &
contemplation.[27]

However, he went on, in reality not all people of good sense were in London,
and the intellectual nourishment on offer could be depressingly meagre; only
rarely did one go home in the evening any the wiser after "the ordinary conversation of the Town." His bolthole at Grantham was a classical retreat from all
the perils of the city, and he related to Gale that he had spent many months the
previous summer completely absorbed in reorganizing his house and garden,
"laying out the stations of dyals, urns & statues, inoculating mistletoe, & trying
vegetable experiments." Stukeley had already built and fitted out a fine study,
which overlooked the garden and the surrounding valley, and which was also
within earshot of a great cascade of the river -- which during the day "raises
the mind to a pitch fit for study." None of this had eroded the possibilities for
learned friendship, and he had already started a weekly meeting for genteel
conversation as well as a "small but well disciplined" Masonic Lodge.[28]

The countryside, at least in this part of Lincolnshire, was – according to
Stukeley -- greatly beneficial to his health. He told Gale that he had worked so
hard in the garden that he had sweated out the London fog and was now able
to eat almost a whole fillet of veal without the digestive support of orange.
Now he was "vastly athletic," and had returned to an older and entirely natural,
druidic form of well-being:

> my antient country complexion is returned to my cheeks, the blood flows
> brisk through every anastomosis, my lips recover their pristine red, &
> my own locks, moderately curled, resemble the Egyptian picture of Orus
> Apollo, or the emblem of rejuveniscence.[29]

Whether in the city or the countryside, or in the liminal places between them,
the garden was central to early eighteenth-century genteel conceptions of
retirement. Indeed, its significance as the key locus of contemplative retreat

---

27   Stukeley to Samuel Gale, 6 February 1726/7; Lukis, *Memoirs*, 1: 188.
28   Stukeley to Samuel Gale, 6 February 1726/7; Lukis, *Memoirs*, 1: 188–90 and ibid., 1: 120 for
     an account of "mental pleasures." See also J.D. Hunt and P. Willis, eds., *The Genius of the
     Place: The English Landscape Garden, 1620–1820* (Cambridge, Mass.: MIT Press, 1988).
29   Stukeley to Samuel Gale, 6 February 1726/7;

was restated continually in the writings of contemporaries such as Alexander Pope. It would ravish Gale, Stukeley told him, to think with what pleasure he was now walking about his garden with a book in hand; these were "my own territorys, mea regna, as Virgil calls it, surrounded with the whole complication of nature's charms." Now, "freed from the hideous crys & nauseous noises of the town", Stukeley was able to return to his writings and drawings of Avebury, and he told Gale that he would publish these within the next couple of years if business did not intervene. Gale replied that, mired in London as he was, he did not have the benefit of such a fabulous location and could only console himself by returning to the "antiquarian periods" (i.e. meetings of the Antiquarian Society) spent discussing and studying the ancients, "preferring thus the few instructive dead to the more numerous & senseless living."[30]

In early April 1727, writing under his Society of Roman Knights persona "Chyndonax", Stukeley waxed lyrical to Samuel Gale about his garden and the views available from his house in Grantham:

> You can't but imagine that traversing a little spot of ones own is vastly more delightful than even the mall, or the heath of hampstead, or the ring, and the like, where we have nothing proper but the common air, & scarcely that without the sophistication, & corruption, necessary to the neighborhood of you Capnopolitans.[31]

He told Samuel – as he had done Roger Gale and Newton before him -- that the house next door had firm foundations and a good orchard, and was available for £200. Stukeley added that he would negotiate on his behalf and would even act as Gale's gardener if he wanted to use it as a summer lodge. The surrounding country was great for hunting, riding and antiquities, and its air and views were unrivalled. All this healthy living was not good for his own trade, he joked, but luckily his practice was going well. He had built up a gigantic store of knowledge in London, and so did not believe he had wasted his time there, but now the time had come to process the "infinity of drawings, & materials of antiquitys, & philosophy" he had put together. Without the leisure to digest this information, study was burdensome: "living always in London is like being at a continual feast, gorging one's self without remission, & overloading the intellects with a confused & distempered medley." He told Gale that the country life would give him the time to digest the fruits of his labour, no longer held

---

30   Samuel Gale to Stukeley, March 30 1727; Lukis, *Memoirs*, 1: 190–1 and 192–3 (from *Eclogues*, 1: 70). For Pope see Mack, *The Garden and the City*, esp. 77–105.

31   Stukeley to Samuel Gale, 3 April 1727;

back by the other bane of his existence – the continual acquisition of new material. However, two years later, in a typical volte-face, he was complaining that he could never complete his antiquarian studies unless he was in London, since only there did he have access to the relevant books.[32]

In October 1727 Stukeley was still entranced by the possibilities of his new existence, and professed complete lack of interest when Samuel Gale mentioned a recent attack on his *Itinerarium*. He told Gale that he had not seen the tome on the grounds that he had stopped buying books, but in any case, he had just completed fitting up his library "(& 'tis just full), so that I may properly say I begin to live." The library – crucial for the learned *dominus* – overlooked a river and a well, and Stukeley added that he had adorned the room with busts, urns, bas reliefs and drawings of Roman antiquities, all matching the Egyptian antiquities he had set up in his adjoining bedroom. Regarding the criticism of his work, Stukeley offered a classical account of his own detachment from the spite of academic pedantry:

> My retreat secures me from malice & envy & all other kinds of paper-gall. I look upon myself as dead to London, & what passes in the learned world. My study is my elysium, where I converse with the immortal ghosts of Virgil, &c., with the old sages & prophets of Egypt, that first disseminated wisdom through the world, & nevertheless, though I be defunct to your side of the world, I revive to a fresh life here; for I fancy myself younger than ever I was, that is, I have a better state of health.[33]

He boasted that he could make an annual salary of £200–300, which was quite enough for his purposes, and that "without too much hurry & fatigue ... for I will never make myself a slave altogether to getting of money, no more than I would to fruitless studys." If money suddenly became harder to come by in Lincolnshire, it would nevertheless last longer than the "overgrown" estates of London physicians, which soon dwindled away – along with their owners. Had he stayed in London, he told Gale, he was sure he would by now be "crammed into one of your hellish vaults under a church"; now he had the chance of being "late laid under a green turf," and incorporated back into the soil.[34]

---

32 Stukeley to Samuel Gale, 3 April 1727; Lukis, *Memoirs*, 1: 194–6; Stukeley to Roger Gale, 22 April, 1729; ibid., 2: 264. 'Capnopolitan' or 'smoky town' comes from the Greek καπνός, 'smoke'.

33 Stukeley to Samuel Gale, 25 October, 1727; Lukis, *Memoirs*, 1: 199.

34 Stukeley to Samuel Gale, 25 October, 1727; Lukis, *Memoirs*, 1: 198–200; Hearne's jaundiced view of Stukeley, is in ibid., 1: 169–70. For the ideal layout and meaning of the scholarly library, see R. Regosin, *The Matter of my Book: Montaigne's Essais as the Book of the Self*

Roger Gale told Stukeley in February 1728 that he hoped that Stukeley's medical commitments would not stop him returning to his "former 'amusements" (i.e. his travels) "in the misfortune of a healthy season." Gale also mentioned Stukeley's recent betrothal to Frances Williamson, which, he said, was almost incredible given Stukeley's previous aversion to the idea of marriage. Luckily, he added, the new Mrs Stukeley, a one-time member of the Society of Roman Knights, was very learned and would be an excellent companion in his studies. Two months later Stukeley told Maurice Johnson that he had finished refurbishing his study, which looked out over his gardens, as well as over the river, heath and pastures. The garden he had designed so that it could be easily traversed on foot – "I think gardens are chiefly designd for walking, and therefore should consist mainly of walks." Stukeley's rectangular garden was composed of 4 parterres, with a crossed gravel walk in the middle, beyond which was a "bason or amphitheatre" 100 feet in diameter. This had a number of different sorts of fruit tree and was the "*sylva academi* for philosophers to walk in." The amphitheatre was surrounded with alcoves and seats, while to the south of the house was his "hermitage vineyard", whose walls were festooned with vines. He contrasted the Lincolnshire scenery with the best of what London could offer, since from the tops of hills close to his house he could see as far as Sherwood Forest and survey the ancient landscape that frequently gave up its Roman antiquities.[35]

Stukeley's retreat from London coincided with his intensified interest in the culture, and particularly the religious settings of the ancient Britons, and his garden was recast as a druidic temple. He told Samuel Gale in October 1728 that his wife had just miscarried for the second time and that he had buried the embryo under the "high altar" in the chapel of his hermitage vineyard (figure 2.2) – this was his "roman" altar, now located in a niche in an ivy-covered wall. Stukeley referred to a circle of tall trees in the form of a giant hedge 70 feet in diameter, around which there was a walk making the entire feature 100 feet in diameter. In the innermost circle (the temple itself), there was an ancient apple tree overgrown with "sacred" mistletoe. Around this was another concentric circle 50 feet in diameter, made up of "pyramidal greens", designed in imitation of the inner circles at Stonehenge (figure 2.3). This, he

_____

    (Berkeley: University of California Press, 1977) and A. Ophir, "A Place of Knowledge Re-Created: the Library of Michel de Montaigne," *Science in Context* 4 (1991): 163–89.

35   Roger Gale to Stukeley, 6 Feb. 1727–8; Lukis, *Memoirs*, 1: 201–2; Stukeley to Johnson, mid-April 1728 in Honeybone and Honeybone, *Correspondence of William Stukeley and Maurice Johnson*, 203–5. See further M. Reeve, "Of Druids, the Gothick and the Origins of Architecture: The Garden Design of William Stukeley (1687–1765)," *The British Art Journal* 13 (Winter, 2012–13): 9–18.

FIGURE 2.2   The Hermitage in Stukeley's south garden at Grantham, june 1727. Bodleian
library, Gough Maps 230, fol. 412.

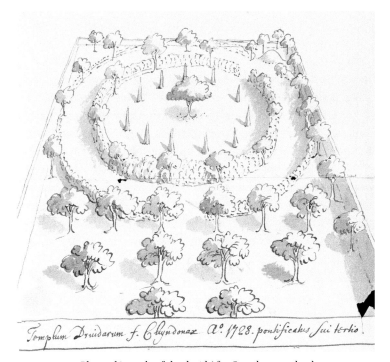

FIGURE 2.3   Planned 'temple of the druids' for Grantham garden by
'Chyndonax', i.e. Stukeley, 1728, Bodleian library Ms. Eng Misc.
c 538, f.10r.

told Gale, was what occupied the time of country folk such as himself, content not with London tittle-tattle "but with nature's converse, where we meet with no envy, slander or uneasiness." A remarkable letter sent soon after this to the freemason and doctor Robert Thomlinson, and written in the guise of a believer in transmigration, shows how Stukeley conceived of his retirement from the metropolis as the liberation of his soul from its "material clog." As such it represented the extinction of his former self – his "death", as Thomlinson had previously termed it. Marking his letter "from Elysium", Stukeley remarked that he had cast off his material form, and was now "in the mansions of peace & felicity."[36]

5   Living and Dying in Stamford

Stukeley's political and religious views inevitably brought him into conflict with the locals. Tories objected to his Whig politics, and there was predictable opposition to his views about the alleged Trinitarian beliefs of the builders of Stonehenge and Avebury. Nevertheless, despite lacking any formal training in divinity, he was encouraged by Archbishop Wake to take holy orders and was ordained by Wake himself in July 1729. In February 1730 he took up the living of All Saints at Stamford in Lincs, telling a correspondent (probably Samuel Gale) at the end of 1729 that many of his friends would marvel at his taking orders, but that he had done so with the purest of intentions. London conversation and "being laughed out of going to church" had contributed to making him talk in a "loose way" but in truth, he said, he had always had religious inclinations. The "sweet tranquillity of country retirement" and "self conversation" in his garden had forced him to look into his own mind, and there he had found the "latent seeds" of faith. Although he noted that various studies had hindered him from examining the most important aspects of religion, his analysis of Druidic antiquities had fuelled his intention to engage in more serious religious study. In time, he continued, he would prove that those "religious philosophers" had an exact notion of the doctrine of the Trinity, and he

---

36   Stukeley to Samuel Gale, 14 October, 1728; Stukeley to Robert Thomlinson, c.1729; Lukis, *Memoirs*, 1: 208–9 and 210; Reeve, "Of Druids," 10–13. For the interplay between Stukeley's garden designs and his appreciations of Stonehenge and Avebury, see D. Haycock, "'A Small Journey Into the Country': William Stukeley and the Formal Landscapes of Avebury and Stonehenge," in M. Aldrich and R-J. Wallis, eds., *Antiquaries & Archaists: the Past in the Past, the Past in the Present* (Reading: Spire, 2009), 46–61.

had been led into the deepest recesses of true divinity by scrutinizing their own ideas more closely.[37]

The reality of Stukeley's rural idyll could not match the extraordinary expectations that he had harboured for it over many years. In May 1729, even before his move to Stamford, Roger Gale had commiserated with Stukeley over his inability to achieve the happiness he had desired from his rural retreat. This was a different sentiment from the one Gale had expressed to Stukeley a few months after the latter had left for Grantham. Then, Gale had written of his envy for Stukeley's contentment in his new setting, and had expressed his hope that he himself would "be blessed with a quiet retreat from the world" before shedding his mortal coil. In June 1729 he told Stukeley that he thought Stukeley had left London too precipitately, but hoped that his new life as a man of the cloth would satisfy his hopes. However, he cautioned that it was not a good idea to publish his theories about the similarities between Druidic and Christian views until they were based on firmer grounds. Stukeley's income from his medical work was also diminishing rapidly, and he himself was dogged by persistent bad health, finding to his dismay that there were peccant elements of the country just as there were in the city. Riding many miles at unseasonable hours in the cold and wet, and lying in damp beds often brought on a violent fit of gout: "the hurry of mind & body wh it threw me into ... discomposed the serenity of ones thoughts, wherein the chief pleasure of life consists." It was partly this, he noted later, that that had prompted him to take orders, though his new salary of £110, which was more than double what he had been earning from his medical practice, was also a key factor. Having experienced the same physical discomfort at Stamford, along with continuing political and religious opposition, he was tempted in the summer of 1730 to pursue efforts to become Secretary of the Royal Society, but he lacked the support necessary for election.[38]

Much later, after his final return to the Capital in 1748, Stukeley recalled that the lack of conversation in Stamford and the rural solitude had been unbearable: "all this while that I lived in the country what I knew was intirely to my self, no one person convers'd with me in that way, or had any regard toward it." While distance from the metropolis and time spent alone remained central to the ideal academic existence, the *kind* of extreme isolation he experienced in the

---

37   Stukeley to (Samuel Gale?) c. November 1729; Lukis, *Memoirs*, 1: 227–8; see also Haycock, *Stukeley*, 189–95 and R. Hutton, "The Religion of William Stukeley," *The Antiquaries Journal* 85 (2005): 381–94.

38   Gale to Stukeley, 7 December 1726, 8 May and 14, 19 and 30 June 1729; Lukis, *Memoirs*, 1: 108–9, 187, 214 and 233–36, and 2: 264 for his disappointing salary at Grantham.

early 1730s was as corrosive of intellectual life as it was distasteful: "I lost the pleasure of a garden, & pasture for horsekeeping, & by degrees found out the great want of literary conversation, without which study is but trifling." During his entire time at Stamford he had only managed to engage in any sort of intellectual discussion with three people. Without "good company & ingenious conversation", he noted, the countryside was an unhealthy forum for study: "the facultys of the mind sink & flag, & at best such an one can be but said to live a dead life there." At Stamford "there was not one person, clergy or lay, that had any taste or love of learning & ingenuity, so that I was actually as much dead in converse as if in a coffin." For the inveterately sociable Stukeley, such loneliness was intolerable, and it was no surprise that his two efforts to found the 'Brazen Nose' literary society (in 1736 and 1745) petered out. In one particularly jaundiced comment on his lot, he remarked that when Great Men retired into the country "where they may better pursue their observations of nature, [they] are so far from being caress'd by the country that they become the objects of their spite and scorn." To this he appended Aulus Gellius's famous observation in *Noctes Atticae* (Bk. 6, §20: 1–3) that when Virgil was prevented by his Nolan neighbours from linking the town brook to his villa, he revenged himself by scratching out the name of the town from his "immortal" *Georgics*, replacing it with a more general reference to the local region (*ora*).[39]

Stukeley was not one to rest on his laurels, and he tried tirelessly to recreate an *urbs in rure*, while striving at the same time to create the perfect conditions for the contemplative life. In 1734 Stukeley purchased a piece of land near his church that he called the "Hermitage" and two years later he bought an adjoining plot. This allowed him both to extend the horticultural ideas he had developed at Grantham, and to create a site for his Brazenose society, which witnessed a number of different scientific experiments during its short existence (figure 2.4). The garden he designed at this property was conceived as a place for perambulation, surprise and contemplation, and as at Grantham, Stukeley embedded a number of druidic features within it. In 1739, two years after the death of his first wife, he married Elizabeth, the sister of the Gales. With what was now a reasonably successful business, an inheritance from the death of Frances, and a £10,000 dowry that accompanied his second marriage, Stukeley could now devote increasing amounts of time to his garden and to his academic work. In the early 1740s he published his great works on Avebury and

---

39   Lukis, *Memoirs*, 1: 77, 104, 108–9, 124–5. Note also Roger Gale's remarks on Stukeley's frequent complaints about rural solitude in ibid., 1: 465ff. Gellius referred to *Georgics* Bk. 2 lines 244–5.

FIGURE 2.4   The Hermitage at Stukeley's house in Austin St. Stamford, c.1738, Bodleian
library, Gough Maps 16, fol. 53 a.

Stonehenge, which were notable as much for their accurate depictions of their
sites as for the idiosyncratic religious uses that Stukeley that ascribed to them.[40]

In March 1741 Stukeley bought a two-acre plot of land in Barn Hill in Stamford, and spent much of the next two years substantially redesigning and
rebuilding the site. At a time when he was preoccupied with visiting druid
temples, and with publishing his work on Avebury, Barn Hill was designed as a
showcase for an ideal place of solitude and retirement. He built a study-library
in an ambitious bridge that linked the main property to a small plot of land on
the other side of the road ('The Street' in figure 2.5) that ran by his house, and
once his house had been completed, the garden, work on which commenced
in 1744, became the main focus of energies. Stukeley added a grotto, a taber-
nacle (or "Temple of Flora") to house Elizabeth's curiosities, and an obligatory
hermitage. The latter was now a building of devotion, constructed out of the
varied remains of medieval churches and monasteries (figure 2.6). In a nod
to contemporary fashion in garden design, he also enclosed a "wilderness"
inside a surrounding walk. Throughout the 1740s, he continued to set up new

40   For Stukeley's gardening and design activities at Stamford see Reeve, "Of druids," 13–16
      and in particular the fine account in J.F. Smith, "William Stukeley in Stamford: His Houses,
      Gardens and a Project for a Palladian Triumphal Arch over Barn Hill," *The Antiquaries
      Journal* 93 (2013): 353–400, esp. 356–67.

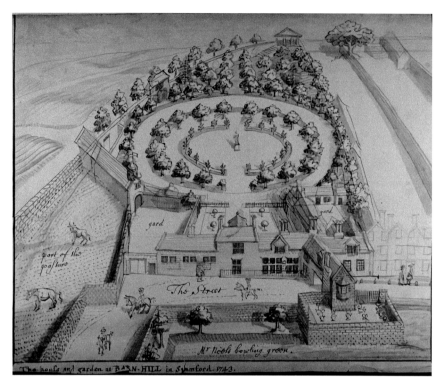

FIGURE 2.5   Stukeley's representation of his house and garden at Barn Hill, 1743, from
Spalding Gentleman's Society Stukeley Portfolio (unfoliated).

FIGURE 2.6   The Hermitage at Barn Hill, 1744, with caption: "Animas quiescendo sapientiores
fieri" ("souls grow wiser by resting"), Bodleian library, Gough Maps 230, fol. 410.

clubs and societies in Stamford and its environs, and occasionally he found the time to explore various landscapes, reporting in detail to the Spalding Society and other local gatherings on the fossils and ancient treasures that he had uncovered.[41]

6   London: Sociable Solitude

In 1739, Stukeley acquired a second living from his patron the Duke of Ancaster, and he became one of the duke's chaplains. For four years (1740–43) he wintered in Gloucester Street in London and participated enthusiastically in ventures such as the newly founded Egyptian Society, although he continued to be based in Stamford for most of the year. In September 1747 he told an old friend from the Egyptian Society, the duke of Montagu, that he had no thoughts of coming to London the following winter, and no mind to ever visit the city again. The following month he told Samuel Gale that he had travelled around Huntingdonshire and Lincolnshire, stopping by the Spalding Society to give a talk on a recently discovered Roman pavement. The man who sent him a drawing of the object had also mentioned the discovery of the body of a woman in Yorkshire, found with tanned skin and leather sandals. There was a surfeit of activity, culture, and curiosities outside of London, and so, he told Gale, "we enjoy the country truly: and that is true life, not the stink and noise and nonsense of London." Nevertheless, Babylon retained its attractions and within weeks of writing to Gale, Stukeley accepted an offer from Montagu to take over the living of St. George the Martyr in Queen's Square, Bloomsbury. Of course, he still believed that the Capital was a hotbed of immorality, and he noted that while one half of the Royal Society were infidels, and the other consisted of fanatics. However, his income from St. George's was greatly superior to what he could derive from the country, and he rejoiced that it would allow him to work in retirement "for God's honor." Most importantly, he wrote, the best thing about living in London was "that I can enjoy what retirement I please, & what company." It was this new found capacity to blend solitude and sociability to the degree that he wanted that was so satisfying, representing the end of a long quest.[42]

---

41   Stukeley to Samuel Gale, 7 June, 1742 and 5 September, 1742; Lukis, *Memoirs*, 1: 335–6; Smith, "Stukeley in Stamford," 381–93.

42   Stukeley to Samuel Gale, 27 October, 1747; Lukis, *Memoirs*, 2: 291–2 and 1: 77, 81, 104, 108–9 and 114–5 (for Montagu).

In the spring of 1748, Stukeley gave Maurice Johnson a number of fossils, scientific instruments and pieces of antiquity for the Spalding Society's Museum. By now he was based in the Capital, and he informed Johnson that he had considered going straight from the weekly meeting of the Royal Society to that of the Society of Antiquaries (which was held on the same day), but was unable to manage both. Rather, his preference was to go back home and write up his recollections of the first meeting over a "contemplative pipe." He told Johnson that he loved the *kind* of solitude that he could experience in London – the beauty of city life was that "we can mix company & solitude, in a just proportion. Whilst in the country, we can have nothing else but solitude." Thus had he found it in Stamford. Three years later he repeated the same sentiments, confessing to Johnson that he no longer had any interest in coffee house culture but now enjoyed the same retirement as he had done in the country. He had recently procured a library by buying an adjoining room to his house in Kentish Town, and was busy fitting it up for scholarly activity. With money, society, and the ability to manage his time and space, Stukeley was free at last to pursue his extended range of interests. He remained an active member of all the major metropolitan intellectual societies, contributing papers on antiquities, genealogy, numismatics, etymology, earthquakes, electricity and astronomy.[43]

Stukeley's accounts of his pursuit of *otium* were highly idiosyncratic, but they are representative of a more general existential problem affecting the life of the intellectual. The scholar must deal with a series of complex problems involving time-and space-management, achieving a proper balance between the lives of engagement and disengagement. A large part of Stukeley's life was spent assessing the relative merits of the active and contemplative lives, each of which he experienced to various degrees. For much of his life, the countryside represented a series of values and locations that offered optimal conditions for pursuing his antiquarian and philosophical interests, but once he was ensconced in rural Lincolnshire he pined for the benefits of sociable conversation. His account of *otium* was steeped in the language and ideas of classical writers, and he repurposed their accounts to make them applicable to eighteenth-century Britain. Like the great Roman authors, Stukeley celebrated the advantages that rural *otium* offered for cultivating virtue, good health, freedom, contemplation, solitude, friendship and useful work, and within this framework he pursued his "curious" interests in gardening, antiquarianism and natural philosophy. Given his belief that the druids practised a sort of proto-Christianity, there is even a sense that his Lincolnshire villas provided an *otium*

---

43   Stukeley to Johnson, 25 May 1748 and 13 April 1751; Honeybone and Honeybone, *Correspondence of William Stukeley and Maurice Johnson*, 121–2 and 167; Lukis, *Memoirs*, 3: 14.

sanctum. However, metropolitan society was the locus for serious intellectual discussion about science and antiquities, and indeed, it was the place where the major learned societies did their business.

The country and the city were more than mere symbols or rhetorical tropes, and Stukeley carved out a variety of lives either in London or in the countryside. He made practical decisions to earn sufficient lucre to sustain his scholarly lifestyle, and he built real villas, libraries and gardens in the countryside. *Otium* was what one made of it, and it had to be forged from what was at hand. In his own pursuit of the "True" or "real Academick life," he found that the city and country *by themselves* could not provide the combination of solitude and company that were essential for academic leisure. To be a real scholar, one had to export elements of the city into the country – by instituting societies that promoted useful conversation – and to import elements of the countryside into the city – by nurturing gardens and protecting the right sort of solitude. In the end, it turned out that the opportunities for creating learned institutions in his beloved Lincolnshire were too limited, and his efforts to institute academic sociability there merely produced a maddening loneliness. Perhaps to his surprise, Stukeley found – as a wealthy clergyman in his early 60s – that only the city could provide the right conditions to balance solitude with company.

CHAPTER 3

Something is in the Air: Experimental Spaces, Analogical Reasoning, and the Problem of Putrefaction in Enlightenment Europe

Margaret Carlyle and Victor D. Boantza

1   Introduction

By the middle of the eighteenth century, the problem of putrefaction in the animal, as well as vegetable and mineral, kingdoms was well known. Across a variety of sites, investigators worked against time, dampness, and heat, the key factors in the decomposition of organic matter. Anatomists hastened to dissect rapidly decomposing corpses, reserving the cooler winter months for their investigations, while seeking out new techniques for drying out body parts for study or preserving them in jars brimming with spirit cocktails.[1] In kitchens, cooks complained of spoiled foodstuffs, while the challenge of preserving them was a concern among others for the pioneers of industrial food production.[2] Surgeons stationed in army garrisons noted the urgency of dressing war wounds, to avoid the corrupting influence of outside air.[3] A central preoccupation for men on the high seas was likewise of a medical nature, as surgeons sought out and administered antiscorbutic antidotes like citrus fruit, in their fight against scurvy, understood at the time as a putrefactive malady.[4] Botanists meanwhile noted that organic matter ironically thrived in putrid waters while alive, yet immediately joined their degenerating ranks upon death.

---

1   On the history of anatomical injection as an early modern European form of preserving putrefying flesh, see F.J. Cole's classic study: "The History of Anatomical Injections," *Studies in the History and Method of Science*, ed. C.J. Singer, volume 2 (Oxford: Oxford University Press, 1921), 285–343.

2   For an account of industrial food production and the relationship between gastronomy and chemistry, see E.C. Spary, *Feeding France: New Sciences of Food, 1760–1815* (Cambridge: Cambridge University Press, 2014), 181–4 (on foodstuff preservation).

3   Such associations and views go back to Renaissance battlefield surgeons. See Lois N. Magner, *A History of Medicine* (New York: Marcel Dekker, 1992), ch. 4, esp. 164–166.

4   See Janet Macdonald, *Feeding Nelson's Navy: The True Story of Food at Sea in the Georgian Era* (London: Frontline Books, 2014).

What was less established than the problem of putrefaction was a sure method for retarding, halting or, even better, undoing its progress, especially acute in the animal realm. By the 1750s, a variety of investigators took up the mantle in the hopes of building a new field of research with practical applications. The likes of English surgeon John Pringle (1707–1782) and French experimenter and scientific translator Marie-Geneviève-Charlotte Thiroux d'Arconville (*née* Darlus, 1720–1805) embarked upon ambitious field and laboratory work aimed at establishing anti-putrid substances, which they identified as "antiseptics," in contradistinction from "septics," which accelerated putrefaction. (figure 3.1) Their contributions were supported by others working along similar lines, including Irish surgeon David Macbride (1726–1778), whose naval services attuned him to the dangers of scurvy. Their separate but complementary work proved foundational, serving as the target of both admiration and critique of the contributors to the prize contest on the subject of "antiseptic substances," held by the Dijon Academy in 1767. What emerged, for instance, from the French surgeon Toussaint Bordenave's (1728–1782) essay—which received honorable mention in the competition—was a critique of the kinds of animal matter used by laboratory investigators, typically the healthy flesh obtained from recently slain cattle, freshly laid eggs, or line-caught fish. Instead, Bordenave urged experimenters to turn their attention to animate bodies.

Many of the investigators who experimented with antiseptics in the period from the middle of the century to the 1770s were motivated by the prospect of developing remedies for army patients suffering from necrosis on the battlefield. Yet, their sites of experiment were at a remove from camp garrisons and distinctly "laboratory" in orientation.[5] Experimental arenas are thus particularly revelatory of how investigators conceptualized the problem of putrefaction and subsequently framed their trials on the antiseptic and septic qualities of various substances. The physical spaces of research and the materials and instruments used within them pointed to the limits of the application of their findings, given the leap required to move from the confines of the laboratory to the complex dynamics of the battlefield. The Dijon Academy's noted essay contributors pointed out the limits of likening fresh animal meat samples to the flesh of living bodies corrupted by gunshot wounds or other putrefactive diseases. Investigators like Thiroux d'Arconville, Pringle, and Macbride had nonetheless found reassurance in the analogical thinking characteristic

---

5   On conceptions of the "laboratory," see Catherine M. Jackson, "The Laboratory," (Chapter 21) in Bernard Lightman, ed., *A Companion to the History of Science*, (Sussex: Wiley Blackwell, 2016), esp. 296–300.

of eighteenth-century science and medicine, by deeming their experimental findings applicable in the admittedly untested grounds of daily life.[6]

FIGURE 3.1   *Left:* Sir John Pringle, engraving by W.H. Mote after Joshua Reynolds, courtesy of the New York public library digital collection, 431012, digitalcollections.nypl.org, Public Domain Mark; *Right:* Marie-Geneviève-Charlotte Thiroux d'Arconville, portrait by Alexander Roslin, 1750, courtesy of Wikimedia Commons, Public Domain Mark.

While this analogical thinking made the experimental laboratory orientation and research findings of Pringle, Thiroux d'Arconville, and Macbride open to critique by the time of the Dijon Academy contest, this essay shows that all three investigators both relied on, and acknowledged the limitations of, such an approach. The power of their respective contributions rested on an imperfect analogy between the inanimate flesh samples tested in the laboratory and the living bodies of patients suffering from various putrefactive maladies like gangrene. They nonetheless believed, to varying extents, that the analogy was

---

6   For the emergence of alternative styles of reasoning (to traditional mechanistic explanations), including the rise to prominence of analogical thinking, especially in medicine and the life sciences during the second half of the eighteenth century, see Peter Hanns Reill, *Vitalizing Nature in the Enlightenment* (Berkeley: University of California Press, 2005); Jessica Riskin, *Science in the Age of Sensibility: The Sentimental Empiricists of the French Enlightenment* (Chicago: University of Chicago Press, 2002); Stephen Gaukroger, *The Natural and the Human: Science and the Shaping of Modernity, 1739–1841* (Oxford: Oxford University Press, 2016).

robust enough to establish productive experimental trials and generate applications with medical potential. For Pringle, a multilayered analogical approach meant that inanimate animal flesh was akin to living animal flesh and environmental factors were linked to putrid diseases. For Thiroux d'Arconville, embracing the notion that an analogy might be drawn between dead and living tissue did not mean that trials conducted were directly translatable into useful applications for the living. Indeed, she suggested that the most potent antiseptics were likely too abrasive to use directly on humans, serving more functionally in their undiluted form as preservation cocktails in the realm of taxidermy and mummification. Of all three, it was perhaps Macbride who was most wary of the analogy between inanimate and animate flesh, between the lifeless test material in the chemical laboratory and the living manifestations of nature's laboratory. And yet he, too, pressed on with experimental trials in the hopes of identifying a comprehensive list of antiseptic substances that could be transported across materials and locations, from the scientific workshop to wounded soldiers on battlefields.

2 Army Medicine and John Pringle's Experimental Program

The cluster of investigators who in the 1750s began to study animal putrefaction experimentally was largely motivated by the prospective medical applications of their findings. The difficulty of the task at hand and "perhaps the impossibility of reaching a conclusive theory" was no reason to shy away from pursuing a subject with such considerable potential utility to humans, as Thiroux d'Arconville remarked in the preface to her *Essai pour servir à l'histoire de la putréfaction* or *Essay Providing a History of Putrefaction* (1766).[7] Investigators believed that antiseptics could be applied to halt the manifestations of putridity, from battlefield flesh wounds to scurvy on board ships. In the case of wounds, a number of surgeons shared in print their experiences of dressing them, pointing to the need to act quickly to avoid the degenerative influence of outside air. Even more effective was the application of substances believed to wield antiseptic powers, which traditionally included wine, walnut leaves, aloes, myrrh, alum, borax, and nitre, alongside more recently suggested ones, like brandy.[8]

---

7   Mme Thiroux d'Arconville, "Preface," *Histoire pour servir à l'histoire de la putréfaction* (Paris: Chez P. Fr. Didot Le Jeune, 1766), xviii.
8   Henry Wellcome, *The Evolution of Antiseptic Surgery: An Historical Sketch of the Use of Antiseptics from the Earliest Times* (London: Burroughs Wellcome, 1910), 48.

As early as 1700, the French surgeon Augustin Belloste (1654–1730) wrote about methods to "dress wounds properly to guard them against the attacks of air", which he viewed as a "powerful obstacle to their healing." Speculating about the material causes of this process, Belloste noted the potential "nitrous, viscous, and ... arsenical parts or qualities" of air.[9] In the early eighteenth century, the military surgeon from Normandy, Guillaume Mauquest de La Motte (1655–1737), emerged as a pioneer in identifying the nature and treatment of gangrene based on his firsthand knowledge of the so-called 'pourriture d'hôpital'—hospital rot—which he attributed to the "corrupted air" of hospitals.[10] He advocated the use of brandy to dress wounds before the onset of mortification, a treatment he combined with tincture of aloes. De La Motte's attempt to assist nature represented a departure from the established Hippocratic axiom of using sword and fire. French military surgeon, Élie Col de Villars (1675–1747), meanwhile recommended immediate dressing to prevent harmful aerial exhalations from reaching fresh wounds.[11]

In the case of scurvy, notable sea commanders of the Georgian period, like English Admiral Horatio Nelson (1758–1805) and Captain James Cook (1728–1779), were motivated to establish alimentary antiscorbutics to stave off what was perceived to be another manifestation of animal corruption. The effects of citrus fruit as a particularly salutary antidote to scurvy have been known since at least the early seventeenth century, and oranges and lemons were actively pursued during resupply stops. These notions later received further empirical support from the investigations of Scottish naval surgeon James Lind (1716–1794), who likewise regarded scurvy as a putrefactive ailment, and conducted controlled experiments in the 1740s to study the dietary intake of sailors in pursuit of a cure. In 1753, Lind published *A Treatise of the Scurvy*, which first went virtually ignored.[12]

---

9   Augustin Belloste, *Le chirurgien d'hôpital enseignant une manière douce et facile pour guérir promptement toute sortes de playes* (Amsterdam: Estienne Roger, 1700), 70–81. See also Jean-Marie Le Minor, "Augustin Belloste (1654–1730), Aspects of the Surgery in Military Hospitals During the Piemont-Savoie War (1686–96)," *Histoire des sciences médicales* 35 (2001) : 317–27 and Jean-Marie Le Minor, "Augustin Belloste (1654–1730), de la chirurgie militaire à la thérapeutique mercurielle," *Revue d'histoire de la pharmacie* 89 (2001) : 369–380.

10   Guillaume Mauquest de La Motte, *Traité complet de chirurgie, contenant des observations & des reflexions sur toutes les maladies chirurgicales, & sur la manière de les traiter.* Second edition. (Paris, 1732), III, 300.

11   Henry Wellcome, *The Evolution of Antiseptic Surgery*, 48–49. On De La Motte see also: O'Halloran, *A Complete Treatise on Gangrene and Sphacelus: With a New Method of Amputation* (London: Paul Vaillant, 1765), xxx.

12   See Macdonald, *Feeding Nelson's Navy*, chapter 7. James Lind, *A Treatise of the Scurvy. In Three Parts, Containing an Inquiry into the Nature, Causes, and Cure, of that Disease* (Edinburgh: Sands, Murray, and Cochran, 1753).

Like other physicians and surgeons before him, Scottish physician and reformer John Pringle understood animal putrefaction as an urgent form of fermentation based on his considerable experience in treating gangrene in a variety of hospital environments. An early advocate of ventilation in hospital wards, he wrote in the preface to his 1764 edition of *Observations on the Diseases of the Army* that "two things induced me to prosecute this subject; the great number of putrid cases that were under my care in hospitals abroad; and the authority of Lord Bacon."[13] The nod to Bacon squares well with Pringle's systematic empirical and experimental approach. His observations *in situ* included British military campaigns in Flanders and Germany (1742–43), Flanders (1744–45), Britain (1745–46), and Dutch-Brabant (1746–48). Among the subjects that interested Pringle were diseases occasioned by climatic factors like heat, cold, and moisture, as well as putrid air, diet, and exercise regimes. The types of diseases he examined included inflammations of organs, such as rheumatism, as well as coughs, dysentery, and "hospital and jayl-fevers." His knowledge of army diseases in living patients was complemented by experimental work in chemistry and anatomical dissections of corpses. Pringle explained in his investigation of fevers, contracted in enclosed spaces, that he had "hitherto ... examined the state of the living body; we shall next consider its appearance after death."[14] In the case of dysentery, Pringle's dissection of human cadavers provided an opportunity to observe the effects of the disease on those who had succumbed to it.[15]

Pringle's *Observations* included a lengthy appendix consisting of seven papers detailing his laboratory experiments, which he had initially presented before the Royal Society between 1750 and 1752, following his 1748 retirement from army service. There, he established a list of "septic" and "antiseptic" substances, the former progressing putrefaction, the latter combating it.[16] In the first account, entitled "Experiments upon Septic and Antiseptic Substances", Pringle justified his experiments on the grounds that, "although an enquiry

---

13   Pringle, "Preface," *Observations on the Diseases of the Army. The Fourth Edition Enlarged* (London: 1764), xiii. The first edition of 1752 was titled *Observations on the Diseases of the Army, In Camp and Garrison.* The second edition appeared in 1753 and a third edition corrected observations from further experience in the camps, in 1754. For Pringle's ideas on putrefaction in medical and political contexts see Erich Weidenhammer, "Patronage and Enlightened Medicine in the Eighteenth-Century British Military: The Rise and Fall of Dr John Pringle, 1707–1782," *Social History of Medicine* 29 (2015): 21–43.
14   Pringle, *Observations* (1764), 308.
15   *Ibid.*, 246.
16   The full title of the section, which spans the vast majority of the appendix, is "Experiments upon Septic and Antiseptic Substances; with Remarks relating to their Use in the Theory of Medicine: in several Papers, read at the Royal Society."

into the manner how bodies are resolved by putrefaction, with the means of accelerating, or preventing that process, has been reckoned not only curious, but useful, yet we find it little prosecuted in an experimental way."[17] In 1752, Pringle received the Copley Medal for this work, the Royal Society's most prestigious award.

Pringle thus undertook experiments to discover "standards, whereby to judge of the septic or antiseptic strength of bodies."[18] While acknowledging the problem of testing antiseptics on inanimate flesh, in the absence of conducting trials on animate bodies, he resorted to animal substances as suitable simulacra for human flesh and fluids. Pringle approached his laboratory tests of antiseptics with a multilayered analogical thinking that helped justify his selection of experimental materials and enhanced the authority of his interpretations. Relying on this analogical framework, he presented, in the first place, inanimate flesh as an appropriate testing ground for putrid diseases of the living body. Second, he reflected on the properties of fibrous and liquid bodily substances, and their interactions with antiseptics. Third, he drew links between environmental factors, including meteorology and water sources, and putrid diseases.

First, Pringle sought to ascertain whether volatile salts—which were a common product of distilled putrefied animal matter—would hasten or retard putrefaction. "As to the effects arising from the internal use of them", he proclaimed, "little can be said, unless the kind of disease were precisely stated." The body, when taken as a clinical whole, was too complex to diagnose, which is why Pringle advocated chemical experimental work: "upon the whole, it will be the fairest *criterion* of the nature of these volatiles, to enquire, whether out of the body, they accelerate or retard putrefaction."[19] Having

> confined my enquiries to such things only, as induce putrefaction out of the body: for as mercury, and certain poisons, which taken into the stomach, or absorbed by the veins, have the effect of septics, I purposely omitted them, as not being able to take in so large a field.[20]

Pringle clearly distinguished between *in vivo* and *in vitro* types of trials, claiming that the latter could be better controlled and deciphered. He typically interpreted his findings by way of a blend of analogy and experimental observation, establishing, for instance, that "volatile alcaline salts not only do not dispose animal substances to putrefaction, out of the body; but even prevent

---

17   Pringle, "Appendix," *Observations* (1764), iii.
18   *Ibid.*, xv.
19   *Ibid.*, vi–viii.
20   *Ibid.*, xli.

it", which is why we "may presume that the same, taken by way of medicine, will *caeteris paribus* prove antiseptic."[21]

Second, he suggested a possible analogy between what he called the fibrous material or "solids" and the liquids or "humours" of the body, whose responses to antiseptics still required further experimentation:

> thus far I have recited my experiments upon flesh, or the fibrous parts of animals: I shall next proceed to shew what effects antiseptics have upon the humours. For tho' from analogy we might conclude, that whatever retards the corruption of the solids, or recovers them after they are tainted, will act similarly upon the fluids; yet, as this does not certainly follow, I judged it necessary to make new trials; which, with some experiments on the promoters of putrefaction, the reverse of the former, shall be offered to the [Royal] Society at another meeting.[22]

Pringle undertook tests on both fibrous flesh from slain cattle and liquids including egg yolks, ox's marrow, urine, gall or bile, and animal and human blood. Most of his human blood samples were selected from patients suffering from putrid disease, as well as healthy human blood, thus blurring the boundary between vigorous and ailing patients.[23]

Third and finally, Pringle linked antiseptics to broader clinical-environmental spaces, arguing that when the atmosphere is loaded with the "effluvia" of stagnating water, meats are quickly tainted and ensuing dysenteries coincide with these fevers.[24] "The most general means of accelerating putrefaction," he explained, "is by heat, moisture, and stagnating air."[25] Such thinking established a connection between environmental conditions and the outbreak of epidemics, a point he previously made in *Observations on the Nature and Cure of Hospital and Jayl-Fevers* (1750) in the context of field hospitals erected near putrid marshlands.[26] This link also prompted him to discuss the aeration of ships as an antidote to scurvy, which he and other contemporaries conceptualized as a putrid disease.[27]

---

21   *Ibid.*, x.
22   *Ibid.*, xxvi. For a general discussion, see Hisao Ishizuka, *Fiber, Medicine, and Culture in the British Enlightenment* (New York: Palgrave Macmillan, 2016).
23   *Ibid., Observations* (1764), 264–65.
24   *Ibid.*, xxiv.
25   *Ibid.*, xxxii.
26   John Pringle, *Observations on the Nature and Cure of Hospital and Jayl-Fevers* (London: A. Millar & D. Wilson, 1750), 10–11.
27   See Jan Golinski, *Science as Public Culture: Chemistry and Enlightenment in Britain, 1760–1820* (Cambridge: Cambridge University Press, 1992), 105–116.

In his presentations to the Royal Society in the early 1750s, Pringle identified quinquina or Peruvian bark—a well-known remedy for malaria—as possessing particularly efficacious antiseptic qualities. The subsequent publication of *Observations* provided an expanded list of antiseptics that included chamomile flowers, camphor, and myrrh.[28] Among the substances he tested for their antiseptic qualities on both solids and liquids were aromatic or astringent substances like resins, gums, pepper, ginger, saffron, mint, rhubarb, ground ivy, mustard, and horseradish.[29] Most of Pringle's tests were aimed at retarding the progression of putrefaction, though he was also motivated by the possibility of undoing decay and effecting the "sweetening [of] several thin pieces of corrupted flesh", a process he had accomplished by applying "affusions of a strong decoction of the Bark."[30]

Pringle's work on antiseptics appeared in French in 1755 as *Observations sur les maladies des armées dans les camps et dans les garnisons*. Pringle's *Observations* functioned as an experimental guide in the preparation of Thiroux d'Arconville's 1766 anonymously published treatise, *Essay Providing a History of Putrefaction* (hereafter *Essay on Putrefaction*).[31] (figure 3.2) She credited Pringle in her prefatory remarks with having opened up a new field of research and justified the necessity of her own investigations on the grounds that "the great preoccupations of Mr. Pringle did not permit him to repeat his experiments ... from which we would be in a position to establish a sure theory. I even dare suggest that he sometimes erred."[32] Thiroux d'Arconville thus proposed to fill the lacuna with novel empirical evidence, while nonetheless respecting the codes of Enlightenment gentlemanly etiquette by deferring to the great authority of her English counterpart whose "superiority ... precludes any rivalry."[33]

3   Mme Thiroux d'Arconville's *Essay on Putrefaction*

Thiroux d'Arconville's investigations into the problem of corporeal putrefaction also came to bear on contemporary theories of fermentation. The famous Dutch chemist Herman Boerhaave (1668–1738) had suggested that only vegetables undergo fermentation in one of two degrees, spirituous and acidic, thus

---

28   Pringle, "Appendix," *Observations* (1764), iii.

29   *Ibid.*, xxi.

30   *Ibid.*, xxiii–xxiv.

31   The work was published as "Par le Traducteur des Leçons de Chymie de M. SHAW, premier Médecin du Roi d'Angeleterre," that is: "By the Translator of the Chemical Lectures of Mr. Shaw, physician to the King of England."

32   Thiroux d'Arconville, "Préface," *Putréfaction*, iii. Note that all translations are our own.

33   *Ibid.*, iv.

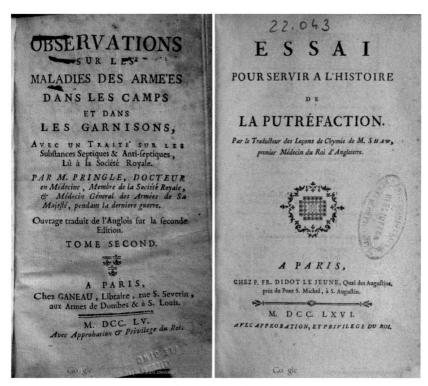

FIGURE 3.2   *Left:* Title page of French translation of John Pringle's *Observations on the Diseases of the Army, In Camp and Garrison*, translated as *Observations sur les maladies des armées dans les camps et les garnisons*, (Paris: Ganeau, 1755), courtesy of the HathiTrust, Public Domain Mark; *Right:* Title page of Thiroux d'Arconville's anonymously published *Essai pour server a l'histoire de la putréfaction* (*Essay on Putrefaction*), (Paris: P. Fr. Didot le Jeune 1766), courtesy of the HathiTrust, Public Domain Mark.

conceptualizing animal decay in a class apart from fermentation. His contemporary, the English clergyman and Newtonian experimenter, Stephen Hales (1677–1761), summarized Boerhaave's viewpoint in 1727 by stating that:

> vegetables alone are the subject of fermentation, but both vegetables and animals of putrefaction ... fermentation is (he says) the motion of the air intercepted between the fluid and viscous parts of the fermenting liquor; but the cause of putrefaction is fire it self, collected or included within the putrefying subject.[34]

---

34   Stephen Hales, *Vegetable Staticks* (1727), 288–89. In 1753, Pringle complained about the "ambiguity" in the use of the words 'putrefaction' and 'fermentation', and wished that "we had two different words to denote the exciting cause of these two intestine motions:

The next generation of investigators challenged this wisdom. The French chemist Pierre-Joseph Macquer (1718–1784) described putrefaction as

> an intestinal movement of fermentation that is excited between the proximate principles of all vegetable and animal matter, from which results a decomposition and a complete transformation in the nature of these principles.

Thus, for Macquer, putrefaction was a veritable form of fermentation. Accordingly, all fermentable animal and vegetable—but not mineral—substances were ultimately prone to putrefaction.[35]

It was his pupil and erstwhile collaborator, Thiroux d'Arconville, who echoed this interpretation, describing matter's innate tendency to putrefy through degrees of fermentation. She likewise summarized this view in a passage that strikingly resembles Hales's words:

> Boerhaave, wishing to provide laws in Chemistry as in Medicine, attributes fermentation itself exclusively to the vegetable kingdom because he recognizes but two degrees: the 'spirituous' and the 'acid.' He excludes putrefaction from this classification, not viewing it as a true form of fermentation, and placing it in a category apart.[36]

Thiroux d'Arconville conceived of putrefaction as a step naturally succeeding Boerhaave's preliminary two degrees of fermentation, and concluded that putrefaction was indeed "true fermentation", the descent into which from fermentation was extremely rapid in much of animal matter.[37] Pringle had advanced a chemico-medical view, stating that "Bacon, as well as some of the Chemists ... hinted at a putrid fermentation, analogous to what is found in vegetables; and this having so near a connexion with contagion."[38]

---

but this is the less to be expected, on account of the disposition of all putrid animal substances to promote both animal putrefaction and a vinous fermentation in vegetables." Pringle, *Observations* 1753, 339–340n. For an overview of the subject see Victor Boantza, "Fermentation," in Dana Jalobeanu and Charles T. Wolfe, eds., *Encyclopedia of Early Modern Philosophy and the Sciences* (Dordrecht: Springer, 2020).

35   Pierre-Joseph Macquer, *Dictionnaire de Chymie, contenant la théorie et la pratique de cette science* (Paris: Chez P. Théophile Barrois, 1778), Volume 3, 282.

36   Thiroux d'Arconville, "Préface," *Putréfaction*, xiv. Note that Macquer was also the anonymous dedicatee of her treatise.

37   *Ibid.*, xiv.

38   Pringle, "Appendix," *Observations* 1764, xxxii.

Thiroux d'Arconville's *Essay on Putrefaction* was recognized for its contribution to the subject as late as 1805, in the *Encyclopédie Méthodique* of Antoine François de Fourcroy (1755–1809), who singled it out as the work of "a French woman ... who distinguished herself by a large number of literary works, most notably the translation of Chemical Lectures by [Peter] Shaw, and who has provided numerous series of experiments on the same subject."[39] Her identification of the antiseptic power of quinquina or Peruvian bark, given its ability to preserve beef samples for as long as seven months and eleven days, squared well with Pringle's findings. She did not reach this conclusion without first systematically replicating Pringle's trials, however, while also expanding the scope of antiseptic substances to include Peruvian balm and metallic salts. The latter nonetheless were deemed toxic and therefore inadequate for the purposes of medical use. Thiroux d'Arconville's aim was to verify Pringle's claims and to examine the applications of her findings in "curing a large number of wounds and illnesses", including those suffered in the battlefield.[40] The resulting *Essay on Putrefaction*—some 600 pages documenting over 300 experiments—formed an attempt to contribute to contemporaries' understanding of how to undo manifestations of putrefaction in both the animal and vegetable realms. The work functioned as an experimental logbook in which the author elaborated on a series of experiments for each of the thirty-two classes of substances she had identified on the basis of their relative antiseptic powers. The text is accompanied by ten recapitulative tables displaying the relative preservation rates of the tested substances (figure 3.3), as well as postscript descriptions of experiments conducted on human bile and ox galls ("fiel de boeuf") extracted from several deceased subjects.[41]

Thiroux d'Arconville culled her experimental data from a series of investigations conducted between 1754 and 1764. Many trials were repeated multiple times for more certain results and to determine points of comparison between similarly powerful antiseptics. She established two separate controls, one with untainted meat, fish, and egg samples left to rot inside a sealed container, and another sealed container for each substance submerged in unaltered water. She

---

39   Antoine François de Fourcroy, *Encyclopédie méthodique ou par ordre de matières: chimie, pharmacie, metallurgie,* Volume 4 (Paris: H. Agasse), 707. On Mme Thiroux d'Arconville's chemical pursuits, see Margaret Carlyle, "Femme de sciences, femme d'esprit: 'le Traducteur des *Leçons de Chymie*'" (Chapter 4) and Brigitte Van Tiggelen, "Entre anonymat et traduction: La carrière d'une femme en sciences" (Chapter 5), in Patrice Bret and Brigitte van Tiggelen eds., *Madame d'Arconville (1720–1805): Une femme de lettres et de sciences au siècle des Lumières* (Paris: Éditions Hermann), 2011, 71–92; 93–107.

40   Thiroux d'Arconville, "Préface," *Putréfaction,* xviii; 442–43.

41   See Thiroux d'Arconville, *Putréfaction,* 499–542 (human and ox bile).

Suc d'Oseille.	1. en Automne.	26 J.	Suc de Ciguë.	1. au Printems.	16 J.	Eau de S. Myon.	1. en Hiver.	11 J.
Extrait d'Aloës.	1. au Printems.	25 J.	Camomille.	5. en Automne.	15 J. 6. J.	Eau de St. Marc.	1. en Hiver.	11 J.
Tacamahaka.	1. en Automne.	25 J.	Cidre.	1. en Eté.	15 J.	Eau de Jode.	1. en Hiver.	11 J.
Sandaraque.	1. en Automne.	25 J.	Eau commune à froid.	5. en Eté.	15 J. 3 J.			
			Idem.	1. en Hiver.	10 J.			
			Idem.	1. au Printems.	3 J.	Ecau de la Madelaine du Mont-d'or.	1. en Hiver.	11 J.
Mastic en larmes.	1. en Automne.	25 J.	idem.	8. en Eté.	15. J. 2 J.			
			Idem.	1. en Automne.	6 J.			
			Au bain marie.	1. en Eté.	9. J.			
Suc de Belle-de-nuit.	1. en Eté.	24 J.	Suc de Scabieufe.	1. en Eté.	14 J.	Eau de la Marguerite du Mont-d'or.	1. en Hiver.	11 J.
Suc de Creffon.	1. en Automne.	24 J.	Suc de Pourpier.	1. en Eté.	14 J.	Eau de Vic en Carladès.	1. en Hiver.	11 J.
Alun.	1. en Automne.	24 J.	Suc de Pavot blanc.	1. au Printems.	14 J.	Eeau des Céleftins.	1. au Printems.	11 J.
Suc de Fenouil.	1. au Printems.	23 J.	Nitre à bafe terreufe.	1. en Eté.	13 J.	Eau de S. Pardoux.	1. au Printems.	11 J.
Gomme Elémi.	1. en Automne.	23 J.	Suc de Solanum	1. au Printems.	12 J.	Lait.	1. en Automne.	11 J.

FIGURE 3.3   Recapitulative table, Thiroux d'Arconville, *Essai pour servir à l'histoire de la putréfaction …* 1766, scanned by the author.

next presented the antiseptic power of a variety of substances applied to her respective animal samples, ordering them from least antiseptic—those preserving samples ranging from a day or two—to most antiseptic, or those staving off putrefaction for many months or even indefinitely. She worked in two different personal laboratories, one was in Paris at 60 rue des Archives (today the Musée de la Chasse) and the other in her country château at Crosne.[42]

Her *Essay on Putrefaction* opens with a description of the Crosne laboratory wherein roughly half of her experiments were conducted:

I was at that time in the country, a few leagues from Paris. The laboratory in which I conducted my experiments was on the ground level. A very large trench, full of live water, bathed one of the walls of this laboratory.

---

42   For more on Mme Thiroux d'Arconville's chemical experiments and laboratories, see Elisabeth Bardez, "Au fil de ses ouvrages anonymes Mme Thiroux d'Arconville, femme de lettres et chimiste éclairée," *Revue d'histoire de pharmacie* 57 (2009): 255–265.

On this same side was a window exposed to the midday sun, and another facing it, consequently exposed towards the North. This site was rather humid.[43]

She later mentioned that this space contained a stove.[44] The layout of her Parisian workshop was somewhat different: "the site in which I chose to conduct my experiments was on the first floor. The room contained nothing but a window exposed to the North and there was no furnace in this room."[45] Thiroux d'Arconville's description of her experimental sites also reveals the set of variables she was interested in charting, most notably the continuously changing elements, as well as fluctuations in sunlight, humidity, and heat.[46]

Thiroux d'Arconville's design to limit experimental variables was hampered by her movement between both sites, an anxiety she expressed in the need to impose some sense of order on the fluctuating empirical landscape:

my experiments were conducted in the city and the country, but to be certain of their exactitude, I never changed locales in either of the two places and I kept a circumstantial Journal to account not only for the state in which I found myself each day, or at least every second day, but also the substances that I used in my experiments and the day, hour, degree of the thermometer, and anything that may relate to the different variations to which the temperature is susceptible in our climate.[47]

The many trials called on a variety of instruments that, much like the sites of production, while indispensable to her work, comprised the source of unwanted irregularities. Her cabinet was complete with Mason jars, strings, filter paper, a double boiler, cloth, and a device for decanting and filtering. With the exception of temperature, which she insisted on measuring using the Réaumur standard, units were localized or employed selectively. We learn of the familiar and foreign: "gros", "lieue", "once" or "pinte" (for water), "livre" for weight, "grains" for salt content, yet no quantitative measure of humidity.[48] She also insisted on applying identical doses of antiseptic substances to flesh

43   Thiroux d'Arconville, *Putréfaction*, 1.

44   *Ibid.*, 13.

45   *Ibid.*, 2.

46   *Ibid.*, "Préface," xxi.

47   *Ibid.*, xxiii.

48   *Ibid.*, xxii n. The appeal of this thermometer as an idealized eighteenth-century scientific instrument is discussed by Jean-François Gauvin, "The Instrument That Never Was: Inventing, Manufacturing, and Branding Réaumur's Thermometer During the Enlightenment," *Annals of Science* 69 (2012): 515–549.

samples of identical weight, to better control results, with time intervals being measured in days, weeks, and months, not hours and minutes.[49] With the aid of blue paper—or occasionally syrup of violets, albeit with less decisive results—she tested the acidity or alkalinity of the medium, in order to compute the delay in putrefaction occasioned by the antiseptic.[50]

Although the physical spaces imposed variations on her experiments, Thiroux d'Arconville constantly eyed the possibility of stabilizing these spaces as controls. Carefully calibrating to counter the instability of apparatus across venues, while also admitting to such fluctuating factors as temperature and humidity, was no mean feat. This subsequently became an exercise in analogy: by comparing the antiseptic power of specific substances in relation to one another, she operated without an absolute constant, but rather with a sliding scale of values that could be seen as internally coherent and whose meaning was gained by self-reference. Awareness of climatic and seasonal change, and the effect of heat on dilation and cold on contraction of animal fibers, often led Thiroux d'Arconville to repeat experiments.[51] As she explained:

> there are some [experiments] that I repeated several times, not only to establish them with greater certainty, but also to compare results with those of other substances which I find comparatively speaking analogous in their [antiseptic] power, but which experiments have nonetheless taught me are very different in their effects.[52]

Substances purporting analogous antiseptic powers, she thus found out, did not manifest themselves identically.

Thiroux d'Arconville may have been more aware than Pringle of the limitations of the animal test subjects—that is, the samples of beef, lamb, fish, eggs, and bile, whose provenance was sketchy at best, but typically recently slain, with her recording on several occasions that the animal had been "killed the previous evening" ("tué de la veille") or "killed very recently" ("tué depuis peu").[53] She also noted that the condition of the sample itself dictated the efficacy of the antiseptic, once again demonstrating an awareness of the effects of seasonal fluctuations of the animal fibers.[54] Substances which facilitated the

---

49   *Ibid.* xxiii.
50   *Ibid.*, xxiv–xxv.
51   On the seasonal dilation and contraction of animal fibers, see Thiroux d'Arconville, *Putré-faction*, 56.
52   *Ibid.*, "Préface," xxi.
53   See examples: *Ibid.*, 1, 10, 219, 245, 355, 359, 463.
54   *Ibid.*, "Préface," xxii.

putrefaction of animal tissues included sugar, gum arabic, and certain salts, whereas metallic salts—corrosive sublimate, blue vitriol, sub-sulphate of mercury, silver vitriol, and mercurial nitre—alongside gums and resins (Peruvian balsam, camphor, burgundy pitch, styrax, and ammonia) were efficacious in staving off corruption. Similarly, certain acids (like wine and vinegar), juices, and quinquina in both extract and powdered form, arrested decomposition of animal matter, assuming pride of place in her list of proven antiseptics.

4   Problems of Application: Antiseptics for Anatomical Preparations

Thiroux d'Arconville concluded that while metallic salts were particularly potent antiseptics, their medical application was imprudent since they were abrasive enough to require dilution or softening even for the purposes of preserving deceased anatomical subjects like birds and insects.[55] Here, the limitations of the analogy between experimental animal subject and its living counterpart is most apparent. Thiroux d'Arconville suggested—without demonstrating experimentally—the medico-surgical application of less abrasive antiseptics, like powdered quinquina and Peruvian balsam. Metallic salts, she indicated, were thus more useful to taxidermists, natural history curators, and anatomists seeking to improve methods of specimen preservation than to physicians and surgeons dressing wounds. By all accounts, Thiroux d'Arconville was tapping into the contemporary fascination with mummies and the revival of ancient embalming techniques, like the account of an Egyptian infant mummy acquired by the Parisian Mme la Veuve Tilliard on rue de la Harpe, described in the Parisian *Gazette de santé* in 1780. Her cabinet of curiosities featured another perfectly preserved subject who could be mistaken for "a sleeping baby" that her (now deceased) husband had acquired some twenty years prior.[56] The foremost Parisian experts in mummification and animal taxidermy were three women. Mme de Grandpré on rue de Prouvaires, Mme Meunier on rue Pastourelle, and Mlle Beaudoin on rue de La Juiverie, all of whom acquired animal corpses that they subsequently embalmed and resold to Parisian, provincial, and foreign collectors.[57]

---

55   *Ibid.*, 442–43.

56   *Gazette de santé* (no. 50, jeudi le 12 décembre 1780), 204.

57   On these women, see: *Journal des dames* (La Haye: Charles Poilly, janvier 1768), 47 and "Art de conserver les animaux," *Avant-coureur* (no. 6 lundi 8 février 1768): 86–88; Guillaume-René Lefébure de Saint-Ildephont, *Etat de médecine, chirurgie, et pharmacie en Europe. Pour l'année 1776, présenté au Roi* (Paris: P.-F. Didot jeune, 1776), 23; and *Journal des Dames* (mai 1766): 114–115.

Across the Channel in London, the *pièce de résistance* of the collection of
the eccentric anatomist, John Sheldon (1752–1808), was reported by Frenchman Barthélemy Faujas de Saint-Fond (1741–1819) following his sojourn in
Britain. Saint-Fond was transfixed by the embalmed body he encountered on
display in Sheldon's home, which he described as:

> a type of mummy, remarkable in two respects; first, by the subject ... and
> second, by the remarkable cares and procedures undertaken in its prepa-
> ration ... the mummy occupies pride of place in the room where the cel-
> ebrated anatomist, who infinitely adores this subject, normally sleeps.[58]

The subject with beautiful brown hair was in fact Sheldon's mistress whom
he had enshrined "in a state of nudity, laid down and sleeping as if on a bed"
on an elegant oblong mahogany table positioned in the center of the room
under a glass frame.[59] The French traveller was shocked by her supple arms,
elastic breasts, and glowing cheeks, concluding that she was in a perfect state
of preservation, her skin retaining something of its color, "despite being in contact with air."[60] Sheldon had prepared the mummy using a laborious process
of injections and skin tanning, followed by removal of organs and viscera for
treatment and replacement in the body, and the assistance of submerging spirits. After the corpse had been prepared, it was sealed for five years in a juniper
box lined with a calcinated chalk designed to absorb humidity.[61]

Instances of remarkably preserved corpses provided occasion to ponder
the nature of putrefaction and its retardants. The Irish-born physician, surgeon, and member of the Royal Navy's Society of Surgeons, James Kirkpatrick
(1696–1770), was surprised upon opening the casket, around 1750, of a man
buried some eight decades before to find an intact body, with supple joints,
healthy complexion, and even more alarmingly, signs of posthumous hair and
nail growth.[62] He observed that:

> as all Putrefaction supposes some previous Looseness, or Rarefaction,
> of the Texture of the putrescent Bodies, which, in Flesh Meat, is evident
> from a preceding Inflation, that terminates in many little honeycomb like

---

58   Barthélemy Faujas de Saint-Fond, *Voyage en Angleterre, en Écosse, et aux Iles Hébrides*
         (Paris: Chez H.J. Jansen, 1797), tome 1, 50.
59   *Ibid.*, 50.
60   *Ibid.*, 50–51. Sheldon was noted for generously sharing "secrets" with his visitors and his
         mummification technique was described at length in *Ibid.*, 51–55.
61   *Ibid.*, 52.
62   For details on Kirkpatrick's life, see Deborah Brunton, "Kirkpatrick, James," *Oxford Dic-
         tionary of National Biography*, online edition (2004).

Orifices; and which, in Vegetables, will rise to a visible Fermentation: And as such Rarefaction necessarily implies the Agency of some Heat, it follows, that such a Degree of Cold, as would totally restrain Rarefaction, must, *a fortiori*, obviate Putrefaction also. For Heat, or the Action of Fire does not more necessarily expand and ratify Bodies, than Cold contracts and consolidates them.[63]

That salts retarded animal putrefaction was well-known, both in the preservation of foodstuffs and of bodies for anatomical investigation or display. Kirkpatrick added in this context that:

> meat newly killed, being carefully dried from its superficial Moisture, wrapped close in dry Linen, and then plunged into a Barrel of good dry Salt, will not only keep sweet for some Weeks in extreme hot Weather, but will also prove as succulent and palatable as fresh, on its being dressed as soon as taken out of the Salt.[64]

Heat and moisture hastened animal putrefaction whereas the dryness of the flesh retarded it. Salt facilitated the latter state by drawing out the moisture from the animal matter. Kirkpatrick reasoned that heat and moisture played a double role as the principles "necessary to the vital Information of Matter" as well as those "requisite to dissolve and unravel the Organization of it." The necessary conditions for the preservation of animal and human life were also those that occasioned the postmortem decomposition of animals and humans.[65] Cold had long since been known to help in blocking or slowing down putrefaction and thus acted as an environmental antiseptic of sorts. But Kirkpatrick also prescribed the use of chemical antiseptics, pointing to the

> Preservation of Flesh by Salt, a Pickle of which, assisted with Alum [potassium aluminium sulfate], is affirm'd to preserve the very Bowels of the human Body from Putrefaction for Years.[66]

Elsewhere he prescribed "the Injection of Brine into Animals newly killed, for the Preservation of their Flesh."[67]

---

63   James Kirkpatrick, *Some Reflections on the Causes and Circumstances, That May Retard Or Prevent the Putrefaction of Dead Bodies* (London: A. Millar, 1751), 7.
64   *Ibid.*, 23.
65   *Ibid.*, 11, 18.
66   *Ibid.*, 21.
67   *Ibid.*, 32.

5   From Mme Thiroux d'Arconville to David Macbride:
       Fixing Antiseptics

Laboratory spaces were an integral part of Thiroux d'Arconville's experimental
trials that, in turn, presented questions about the portability of her findings
to new arenas: did living flesh react to antiseptics in the same manner as her
meat samples? Could antiseptics be diluted for effective treatment of humans
suffering from putrefactive maladies? These were not the only questions
weighing on Thiroux d'Arconville's mind as she prepared her manuscript for
publication. Besides reporting on her hefty experimental enterprise in *Essay
on Putrefaction*, Thiroux d'Arconville ventured with some discomfort into the
world of causation, in the form of chapter-by-chapter concluding theoretical "Observations." She attributed the antiseptic and conserving agent to its
protective capacity against the corrupting influence of contact with outside
air. Hence to prevent putrefaction in the first place, it was necessary to avoid
contact with air. This was in contrast to the findings of her contemporary, the
former naval surgeon David Macbride, who published his *Experimental Essays*
(1764) while Thiroux d'Arconville was completing her own experiments. Her
intellectual honesty prompted her to credit Macbride's work in her "Supplément," contained in the conclusion of her *Essay*. Macbride suggested another
explanation. He believed to have found in fixed air, first isolated by Joseph
Black (1728–1799) in the 1750s, the agent responsible for antiseptic power.
Thus he concluded that all bodies in nature owed their firmness and cohesion of parts to the fixed air they contained; in depriving them from this air,
the adhesive power is weakened, leading to the putrefactive decomposition of
the substance. In the body, this cementing principle bound tissues together
and its loss led to putrefactive processes and illness, which antiseptics could
correct.[68]

What was particularly intriguing for Thiroux d'Arconville in light of her own
experiments was Macbride's correlation of fixed air with the much-debated
phlogiston:

> Macbride's system of *fixed air* gives birth to a new one in my mind on
> *phlogiston*. One knows that in depriving metals of it, they are reduced
> to lime, and in giving it back, they regain their original form, and all the
> advantages they had lost. Can we not then conclude that *phlogiston* is to
> metals what *fixed air* is to the other kingdoms?[69]

---

68   Thiroux d'Arconville, *Putréfaction*, 542–547.
69   *Ibid.*, 546.

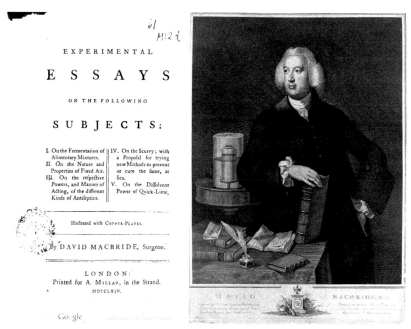

FIGURE 3.4   *Right:* David Macbride, engraving by J.T. Smith, after Reynolds, 1797, courtesy
of the Wellcome Collection, 6066i, Public Domain Mark; *Left:* Title page of
Macbride's *Experimental Essays* (London: A. Millar, 1764), courtesy of the
HathiTrust, Public Domain Mark.

While Thiroux d'Arconville's and Macbride's explanations were inverted—she
focused on the power to guard against the corruption of outside air while he
focused on preventing fixed air from escaping matter—they ultimately con-
founded the brand of putrefaction undergone by cutaneous scurvy with that
involved in gangrenous war wounds.

Macbride was a Northern-Irish surgeon who served in the Royal Navy
aboard a hospital ship and later as a surgeon during the war of the Austrian
Succession (1740–1748). In 1751, he established a practice in Dublin. Given his
naval background, he became especially interested in the treatment and prevention of scurvy, about which he wrote in his *Essays*.[70] (figure 3.4) He saw
himself as working in the burgeoning tradition of pneumatic chemistry, indicating that he was writing a sequel to the illustrious contributions of Black,
Pringle, and Hales.[71] As Jan Golinski has suggested, Macbride was popularizing

---

70   E. L. Scott, "The 'Macbridean Doctrine' of Air: An Eighteenth-Century Explanation of
Some Biochemical Processes, Including Photosynthesis," *Ambix* 17 (1970): 43–57, on 46.

71   David Macbride, "The Preface," *Experimental Essays on the Following Subjects* ... (London:
A. Millar, 1764), xii–xiii.

Black's work by linking it to the environmental causes of putrid diseases. When Pringle became President of the Royal Society from 1772 to 1778, and personal physician to King George III, he advocated the adoption of Macbride's theory of putrefaction while orchestrating the conferral of the Copley Medal to Joseph Priestley in 1773 for his early contributions to pneumatic chemistry, and especially his pioneering work on "different kinds of air."[72] It is worth noting in this connection that Antoine-Laurent Lavoisier (1743–1794) was charged by the Paris Royal Academy of Sciences to investigate the possibility of treating scurvy using the recently identified fixed air.[73]

At the outset of his *Essays*, Macbride stated their aim

> to shew, that there is another principle in matter beside those which are commonly received; and that it is upon this principle, forming the cement, or bond of union, that the firmness, soundness, and perfect cohesion of bodies, chiefly depends.[74]

He drew an analogy between digestion, fermentation, and putrefaction, the link between them being the seemingly spontaneous intestine motion of matter. According to Macbride, this is how matter was being composed, decomposed, and recomposed in various combinations and mixtures in nature. The reference here is to a process of deep-level essential chemical change and not merely a mechanical reordering of the parts of homogeneous matter. Macbride put Pringle's claims to experimental test, using bread, mutton, spinach, and watercress, among other substances. He noted that "these mixtures, being put into phials not closely stopped, were all placed in a moderate degree of heat, on the top of a sand furnace, wherein a retort was at work, on a process which required a continual fire for three to four days."[75] He measured the time it took for each sample to start fermenting and putrefying, and for how long, while paying particular attention to the smell and look of the samples. In line with Boerhaave, Macbride labelled the "stages of fermentation ... [as] *sweet, sour, and putrid*", a division corresponding to the stages known as "*vinous, spiritous,*

---

72   Golinski, *Science as Public Culture*, 107–8.

73   It is also important to note that Joseph Black's "discovery" of fixed air in 1755 in Edinburgh was hardly known in France at this time. On Lavoisier, see Jean-Pierre Poirier, *Histoire des femmes de science en France: du Moyen-Âge à la Révolution* (Paris: Pygmalion, 2003), 269. On Black and fixed air, see the two classic articles by Henry Guerlac, "Joseph Black and Fixed Air a Bicentenary Retrospective, with Some New or Little Known Material," *Isis* 48 (1957): 124–151 and "Joseph Black and Fixed Air: Part II," *Isis* 48 (1957): 433–456.

74   Macbride, *Experimental Essays*, "The Preface," vii.

75   *Ibid.*, 1–4, 5.

and *acetous*."[76] He proceeded to describe the various progressive stages of human digestion, concluding that its main facilitator—and by analogy, a key element in putrefactive processes more generally—"appears to be chiefly the *Fixed Air* of the alimentary substances."[77]

That air existed in some form in bodies has been known since at least Boyle, and it was to Hales that Macbride attributed "the discovery that this elastic matter so nearly resembling common air, is the *principle* which *forms* the *cement*, or *bond of union*, between the several constituent particles" of matter. The principle of cohesion, he reasoned, was therefore air, and not earth, which had been traditionally associated with properties like solidity, firmness, and material coherence. This view also supported ideas related to the economy of nature. Dead bodies, being useless, had to decompose and dissolve, a process that started as soon as the fixed air began flying away from them, allowing putrefaction to ensue.[78]

Macbride next attempted to ascertain "the relative quantities of air, set free from different mixtures by fermentation."[79] To this end, he fermented bread, mutton, beef, lemon juice, sugar, and honey, among other substances, using saliva and bile (ox gall) as fermentative agents.[80] "By this time", he remarked, "I had sufficiently satisfied myself with respect to the manner in which digestion is carried on in the human body; being now fully convinced that it is neither more nor less than a true fermentatory process," adding that the air given off during this process did not cause discomfort due to distension because it had enough space to expand within the digestive tract.[81] Following further experiments on human blood, human bile, and eggs, Macbride argued that:

> The common notion concerning putrefaction, which is universally taught, and as generally believed, is, that bodies become putrid because that *air* hath *access to them*, and communicates somewhat; and few people seem to have any *idea* that putrefaction ensues in consequence of the *loss* of some principle; which, however, appears to be the real cause. For it will be shewn hereafter, that the methods to preserve bodies from putrefaction and decay depend, almost in every instance, on *restraining flight* of the *fixed air*; for, as this principle *cements* and *binds together* the

---

76   *Ibid.*, 8–9.
77   *Ibid.*, 23.
78   *Ibid.*, 28, 30–31.
79   *Ibid.*, 33.
80   *Ibid.*, 53–54.
81   *Ibid.*, 59, 63.

constituent particles of bodies, rottenness, or putrefaction, which consists in the *resolution* and *disunion* of these particles, will not take place
while the *cementing principle* is present.[82]

Macbride concluded that "whatsoever hath the power to *restrain* the
flight of this element [fixed air], or *hinder* the *intestine motion*, will *prevent*
putrefaction."[83] Like Pringle, Macbride asserted that all salts have antiseptic
powers, and would thus be crucial to the "methods to preserve bodies from
putrefaction." Calcareous earths, saline bodies, and astringent substances all
have a strong affinity to fixed air and would therefore retard putrefaction by
preventing its flight.[84]

At the same time, however, Macbride wondered about the parameters of
the *in vitro–in vivo* analogy. How much can be reasonably inferred, on the basis
of experimental trials on antiseptics, about the living body, a considerably
more complex and dynamic system? "It may be demanded", Macbride asked
rhetorically,

> what can these experiments prove with regard to the restoration of putrid
> fluids, in a living body; is it possible to saturate these humours with such
> a quantity of air as will be sufficient to correct their *sharpness*, *restore*
> their *consistence*, and *bring back* their *sweetness*?[85]

Macbride's reflections and intimations about these theoretical and practical
tensions betray a noticeable measure of caution and skepticism, likely trace-
able to his years of personal involvement in medical practice.

In an exemplary instance of controlled and comparative experimentation,
Macbride placed a piece of mutton in a phial with water, which he placed in a
warm environment, letting it stand and putrefy over four days. He then divided
the rotting meat into five samples. On four samples he poured four different
acids—spirit of vitriol (sulfuric acid), spirit of sea-salt (hydrochloric acid),
vinegar (acetic acid), and lemon juice (citric acid), all diluted to roughly the
same strength—while placing the fifth sample back in water. Macbride proceeded to carefully observe the action of the acids over four days, concluding
that, when compared to the water sample that he employed as a standard, he
could confirm the capacity of the acids to retard and reverse the putrefactive

---

82   *Ibid.*, 74–75.
83   *Ibid.*, 120.
84   *Ibid.*, 111, 120–21, 124.
85   *Ibid.*, 149.

processes (figure 3.5). And yet, he observed, "since *acids* both *resist* and *correct* putrefaction, it was very reasonable to expect that all putrid diseases should yield to them, when given in the way of medicine," although medical experience indicated that "their power in this respect is pretty much limited."[86]

Whereas acids worked well within the confines of the chemical-experimental space, they fell short when applied in the context of the medico-clinical space. Reasoning by way of analogy from dead meat to living flesh proved to be a limited exercise. Even though the mechanism and root cause seemed to be identical—Macbride noted that attending to "the things that *prevent* putrefaction in *living* bodies, we shall still find that the dependence is on the quantity of *air*"—drawing analogies between the chemical laboratory and the laboratory of nature did not suffice.[87] The same doubts applied to the antiseptic scope of alkaline substances. Although it was not clear

> whether *alcalies* do in reality promote putrefaction in *living bodies*; there can be no doubt of their power to resist and correct putrefaction in *dead bodies*; but whether, upon the presumption of this virtue, they can be given with propriety, as antiseptics, is not so clear.[88]

In the clinical context, Macbride reverted to his empirical observations and experiences as a practicing surgeon. For the prevention of systematic putrefactive ailments, such as "scurvies, fevers, and dysenteries, to which seamen, and people pent up in garrisons, are often subject", he still prescribed the dietary intake of vegetables, a component of our nourishment that was richest in air that "enters the composition of the animal fluids." Vegetables, then, supplied most air or most "*antiseptic vapour*" to the body.[89] The analogy, however, between the putrefaction of animal substances and the fermentation of vegetable matters was helpful. As both processes were considered to be air-releasing, the ingestion of vegetables, which would ferment within the body, would compensate for the body's loss of air as part of putrefactive disorders. Macbride therefore concluded that scurvy and putrid diseases could be cured by the ingestion of "easily fermentable substances" like vegetable matter, which would soon render into the blood a great quantity of air.[90]

---

86   *Ibid.*, 127.
87   *Ibid.*, 158.
88   *Ibid.*, 151.
89   *Ibid.*, 159–60.
90   *Ibid.*, 176.

Table III. ACIDS diluted tried as ANTISEPTICS.				
A C I D S.	AFTER STANDING			
	24 Hours.	48 Hours.	3 Days.	4 Days.
(1) of Vitriol.	Sweet.	Sweet.	Sweet.	Sweet.
(2) of Sea-falt.	Sweet.	Sweet.	Sweet.	Sweet.
(3) of Tartar.	Sweet.	Sweet.	Beginning to putrify.	Putrid thrown out.
(4) of Vinegar.	Sweet.	Sweet, and much fwelled.	Sweet.	Sweet.
(5) of Lemons.	Sweet.	Sweet, and much fwelled.	Sweet.	Sweet.
(6) Water, as a Standard.	Smell grown offenfive.	Very fetid.	Putrid, and foft.	

POWERS OF ANTISEPTICS.   113

FIGURE 3.5   Comparative antiseptic potency of acids over time, Table III. Acids diluted tried as
             Antiseptics, from Macbride, *Experimental Essays on the following subjects ...* 1764, 113.

6   Animating Putrefaction at the Dijon Academy

The reinvigoration of the problem of putrefaction by the likes of Macbride,
Thiroux d'Arconville, and Pringle prompted the Dijon Academy of Sciences,
Arts, and Literature (est. 1725) to hold an essay contest on the subject of antiseptics. The Academy pointed out that "there are few maladies more common than those controlled by putridity; there are few of which there are so
many kinds; but there are few for which the treatments have been until now so
uncertain." The utilitarian dimension was prominent as the prize contest was
designed to identify antiseptics useful for treating ailments in practical medicine. The Academy emphasized the interdependence of reasoning and observation in the characterization and diagnosis of diseases and remedies. The
call was for essay entries designed "to determine antiseptic substances in the
broadest sense" with a four-pronged goal of: (1) broadly identifying antiseptic
substances; (2) explaining their way of action; (3) distinguishing their different kinds; and (4) indicating their application to illness.[91] In this context, the

91   Barthélemy-Camille de Boissieu, *Dissertation sur les antiseptiques qui ont concouru pour
      le Prix proposé, par l'Académie des Sciences Arts & Belles Lettres de Dijon en 1767 dont la
      premiere a remporté le Prix & dont les deux autres ont partagé l'Accessit* (Dijon; Paris: Chez
      François des Ventes; Chez Des Ventes de la Doué, 1769), v, viii–viii.

Academy singled out the works of Pringle, the anonymous author of the *Essay on Putrefaction* (Thiroux d'Arconville), and Macbride for praise, even though they remained unconvinced that any of these authors had in fact formulated a satisfactory doctrine or a practicable system. All the experiments and observations of these authors to date, the Academy noted, were akin to "diamonds still covered in their sandy shell." Or, in other words, diamonds in the rough.[92]

The Academy received many entries and published the top three essays under the title *Dissertations sur les antiseptiques* (*Dissertations on antiseptics*) in 1769, two years after the contest. The physician at the Montpellier Faculty of Medicine and medical teacher at Lyon, Barthélémy Camille de Boissieu (1734–1770), was declared winner. (figure 3.6) The two runners-up, whose essays received an honorable mention, were Toussaint Bordenave, celebrated master surgeon in Paris and professor at the Royal Academy of Surgery, and Guillaume-Lambert Godart (1721–1794), physician at Verviers near Liège.[93]

Drawing on the findings of Hales, Pringle, Thiroux d'Arconville, and Macbride, Boissieu began his essay by relating a series of experiments on the putrefaction of meat, noting in particular the release of air during putrefaction, as well as the role of external air in facilitating it.[94] Boissieu defined "rot, putrefaction, and putrid fermentation" as "an intestine motion that arises by itself between the insensible parts of an organized body ... and gradually destructing this body by reducing it into its elements." Following this definition, he distinguished between four degrees of putrefaction: first, tendency to putrefy; second, commencing putrefaction; third, advanced putrefaction; and fourth, perfect or completed putrefaction. During the first stage, the animal substance—"the only one which is under question", as Boissieu put it—begins to emit a light smell and some air. If enclosed in a vessel it will start to soften, and if exposed to air its surface will slightly dry out. In the second stage, it becomes acidic, suffers weight loss, acquires a fetid and disagreeable smell, and begins to change its color, volume, and consistency. The third stage brings about the exhalation of "a fetid, insupportable, and nauseating odor" alongside the onset of dissolution and the substantial loss of volume and weight. In the fourth and final stage, the odor diminishes, while the substance acquires a gelatinous texture; finally, it "turns into a crumbly, earthy matter."[95] By way of analogy, Boissieu claimed that there were also four degrees of putrefaction or rot in living

---

92   *Ibid.*, ix.
93   *Ibid.*, x. On Godart, see G. R. Demarée, "Guillaume Lambert Godart: Médecin, philosophe et météorologiste. Un savant oublié du XVIIIe siècle," *Ciel et Terre* 109 (1993): 47–51.
94   *Ibid.*, 5–16.
95   *Ibid.*, 18–19.

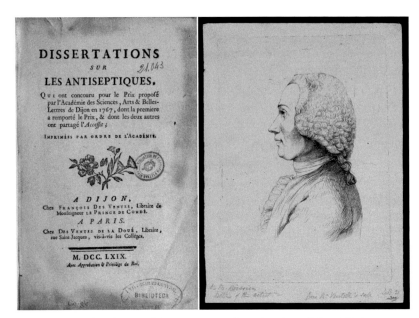

FIGURE 3.6   *Left:* Title page of *Dissertations sur les antiseptiques* (Dijon & Paris:
Francois des Ventes, Des Ventes de la Doué, 1769), the leading essays in
the 1667 Dijon Academy contest on the question of antiseptics, courtesy
of the HathiTrust, Public Domain Mark; *Right:* the winner of the contest,
physician Barthélémy Camille de Boissieu, etching by J. J. de Boissieu, 1771,
courtesy of the Wellcome Collection, 2029307i, Public Domain Mark.

animals, which he labelled in a similar fashion, identifying the third stage, of
"advanced putrefaction", also as gangrene.[96]

Boissieu distinguished between simple antiseptics, meant to be used on
dead animal matter, and medicinal antiseptics, devised to arrest and correct
putrefaction in living beings.[97] The latter, in contrast to the former, "act on a
live animal, in which the solids have an energy and an active force, the fluids
a progressive motion and a reaction." In fact, Boissieu saw a close connection
between life and putrefaction, stating that "all parts of animals tend continually toward putrefaction (incontestable truth). Animalization, or sanguinifica-
tion, can operate only by the action of a putrefying intestine movement."[98] The
reference here is to the longstanding view that digestion and the transformation of food and nourishment into flesh is a form of digestive fermentation.
Like Macbride, Boissieu thought that during putrefaction the body lost fixed

---

96   *Ibid.*, 60–61.
97   *Ibid.*, 27.
98   *Ibid.*, 36–7.

air. As the "cement, the bond between all parts, that gives force to our solids, and consistency to our fluids, [fixed air] tends to continually escape and dissipate through our various excretions."[99] He divided medicinal antiseptics into internal and external.[100] The external ones acted by inhibiting the loss of fixed air.[101] As he reached the part in which he sought to "explain the manner of action of antiseptics", Boissieu argued that "the best means of resolving this part of the problem, is to examine first of all their effect on animal substances deprived of life."[102]

Like Boissieu, Bordenave emphasized that "the decay of animate bodies should especially command our attention." Bordenave conceptualized putrefaction as "the final stage of existence proper to each individual" before it changes its form and ceases to exist as such."[103] He criticized authors, like Pringle and Thiroux d'Arconville,

> who had preoccupied themselves with research into these remedies and who have explained their action, and founded their doctrine on the effects they have observed, by testing animal flesh with diverse substances. These experiments, however illuminating, are illusory in certain respects, or at least, insufficient.[104]

Bordenave thereby questioned the laboratory focus on uncorrupted fresh meat on the grounds that its action "can hardly be compared to that in afflicted parts."[105] He likewise doubted the analytical power of drawing an analogy between living and dead bodies in evaluating antiseptics, since even the most accredited antiseptics might prove ineffective when applied to putrefying wounds.

The third-place entry by Godard on *Septicologie ou dissertation sur les antiseptiques* (*Septicology or Dissertation on Antiseptics*) argued that since heat occasioned fermentation, antiseptics needed to work against it by refreshing (cooling), ventilating, and fortifying bonds in matter, to counter putrefaction.[106] Even more so than his competitors, and as his essay title suggested, Godard's inquiry was heavily focused on the different species, degrees, and causes of

---

99   *Ibid.*, 37–8.
100   *Ibid.*, 39.
101   *Ibid.*, 41.
102   *Ibid.*, 27.
103   *Ibid.*, 142.
104   *Ibid.*, 178–9.
105   *Ibid.*, 179.
106   *Ibid.*, 319.

putrefaction "to which living bodies are susceptible."[107] If anything, he and the other two contributors were calling for a return to the concerns of military surgeons, noting that recent laboratory work on slain meat had reached its limits, especially in the spaces of applied medicine.

All three prize entrants either directly critiqued—or noted by their emphases—the experimental reliance on the freshly slain meat of healthy animals without any putrefaction. This observation is not surprising given the remit of the contest—to test out antiseptics—or indeed, given the common medical background of the three authors. Although Pringle, Macbride, and Thiroux d'Arconville also had their hands in medicine and set their sights on stalling the putrefaction of battle-inflicted wounds, their actual experimental spaces were distinctively laboratory, not battled-field based. If anything, the Dijon contest pointed to the peculiarities of the sites of experiment: the laboratory, neatly captured and documented over time, did not lend to testing the living body and all its variables. Spaces were a limiting factor in a second sense: the containers themselves functioned as sites of experiment, as the presence or absence of fixed air was understood to either retard or accelerate putrefaction. Flesh in the container behaved in one fashion; bodily flesh did in another.

7   The Economy of Nature: Likening the Living to the Dead

Experimental sites were not only the spaces in which trials unfolded, but capsules for thinking that delimited the subjects, materials, and methods of apprehending the phenomenon of animal putrefaction. While laboratories were perfect sites for testing on inanimate animal substances, the hospital and battlefield offered living patients. Analogical thinking, it seemed, ensured the portability of findings across experimental sites. But analogical thinking was not without its problems. On the one hand, an assumption of the fairly uncomplicated analogy between recently slain animal flesh and living matter had emboldened experimenters like Pringle, Thiroux d'Arconville, and Macbride to extrapolate their findings to fresh wounds encountered by military surgeons on the battlefield. The absence of animation presented a problem, however, for contributors to the Dijon Academy's prize contest. Thus one of the limitations of analogical thinking was the discrepancy between site and source material— the living versus the dead; in the battlefield versus the laboratory.

---

107   *Ibid.*, 351.

Analogies functioned across experimental spaces in ways that raised as many problems as solutions. They also operated conceptually, by linking the mineral and vegetable forms of putrefaction to the animal one. The problem of animal putrefaction assumed dominance in experimental trials of the 1750s and 1760s in no small part because of its applicability to human bodies, but also due to a longstanding notion of the superiority of the animal kingdom.

The application of laboratory experimentation to patients reflected an impulse to promote the utilitarian dimension of chemistry. Those who investigated animal putrefaction did so with a view to providing medical practitioners with an arsenal to cure ailing patients *in situ*. While the aim of the findings was borne out of medical utility, they provided more demonstrable potential in the realm of preservation, as Thiroux d'Arconville found, in helping maintain already deceased animal parts for anatomical observation and teaching. The research direction of the Dijon Academy prize contest evinced an interest in bridging that gap, by emphasizing the importance of uncovering antiseptics that were proven to work on living, rather than dead, bodies.

The link between putrefaction and animate bodies was not lost on Thiroux d'Arconville, even if she and her contemporaries carried out their experiments on meat samples that were easier to manipulate and interpret. If the conclusions drawn from her experiments on animal flesh left some of nature's "secrets" uncovered, she reassured her readers that nature had ultimately ordered it this way, right down to the very process of human ageing, the ultimate exercise in the putrefaction of bodily flesh. "We can thus understand putrefaction as the will of nature", she proclaimed,

> and the two degrees of fermentation that precede it as its precursors. Having just reached the age of puberty, and acquiring its strength, does a child lose the delicateness of its features and the freshness of its complexion.[108]

Even though the act of living was a constant process of putrefactive advance that no experimental program might halt, that did not mean there was no value in trying to mitigate and restore the ravages of putrefaction.

---

108   Thiroux d'Arconville, "Préface," *Putréfaction*, xi.

CHAPTER 4

Instrument Makers, Shops, and Expertise in Eighteenth-Century London

Jasmine Kilburn-Toppin

This chapter explores the living, working, and commercial spaces of London's instrument makers across the long eighteenth century. An unlikely body of sources perhaps, the records of London's principal criminal court give insights into a heterogeneous, and highly significant occupational group in England's metropolis. This discussion focuses upon the organisation of space within households and businesses, the typical working conditions and social relationships within workshops, and methods of display within artisan-retailers' shops. This chapter also attempts to uncover the broad outlines of languages of expertise, knowledge, and ownership employed by retailers, craftsmen, and journeymen in these multifunctional sites. I take here a purposefully broad definition of instrument makers, to include mathematical, optical, philosophical, and surgical instruments, and also evidence about devices like clocks and globes.

Pioneering works by E.G.R. Turner, M.A. Crawforth, D.J. Bryden, and Gloria Clifton, among others, have outlined the key features of the booming trade in optical, mathematical, and philosophical instruments in early modern London.[1] The overall narrative of change is now a familiar story. In sixteenth-century London, aristocratic and royal desire for instruments was met by importing objects and skilled immigrant workers from Continental Europe; but by the eighteenth century, the city had a growing international reputation for the quality of its instruments, and the high specialisation and innovation of its makers. The manufacture and sale of instruments was spread across eighteenth-century London, with the most successful makers of optical and

---

1 E.G.R. Turner, *The Mathematical Practitioners of Hanoverian England, 1714–1840* (Cambridge: Cambridge University Press, 1966); M.A. Crawforth, "Instrument Makers in the London Guilds," *Annals of Science* 44 (1987): 319–77; D.J. Bryden, "Evidence from Advertising for Mathematical Instrument Making in London, 1556–171," *Annals of Science* 49 (1992): 301–336; Gloria Clifton, *Directory of British Scientific Instrument Makers 1550–1851* (London: Philip Wilson Publishers, 1995).

© JASMINE KILBURN-TOPPIN, 2022 | DOI:10.1163/9789004501225_005

philosophical instruments lining "the important retail corridor of Fleet Street, Ludgate, and St Paul's Churchyard", while mathematical instrument makers predominated in the eastern side of the metropolis.[2]

Beyond this broad outline of the geography of the trade, we know precious little about the urban spaces within which instruments were made, marketed, and sold. In a discussion of wealthy foreign travellers shopping for instruments in seventeenth and eighteenth-century London, Jim Bennett raised the intriguing prospect that the shops of renowned makers operated not simply as commercial sites, but also as spaces for the exchange of knowledge and skills.[3] Using a very different body of sources, and engaging with a much wider social profile of maker and consumer, this chapter aims to expand our knowledge and understanding about spaces of scientific instrument manufacture and retail. Conceptually it frames 'space' as an active framework, simultaneously produced by spatial practices (such as work, food preparation, and sleep), representations of space (such as maps and plans), and representational spaces (including symbolic associations, like customary links to skill and innovation).[4] I also pay close attention to the materiality of physical space in influencing human behaviour and identity.[5] We thus elucidate the material, social, and symbolic meanings of diverse artisanal spaces in order to deepen our understanding of the culture and society of instrument makers in eighteenth-century London.

The primary sources here include the Proceedings of the Old Bailey, and inventories from the City of London's Court of Orphans. Certainly, these are archives which are more familiar to social or cultural historians of the metropolis, than scholars of the history of science. Testimony from the court room – given by instrument makers as prosecutors, witnesses, and defendants – offers rare insights into the workings and complexities of the trade. In the process of giving criminal evidence, instrument makers, and those living in their

---

2 Alexi Baker, "Symbiosis and Style: The Production, Sale and Purchase of Instruments in the Luxury Markets of Eighteenth-Century London," in A.D. Morrison-Low, Sara J. Schechner, and Paolo Brenni, eds., *How Scientific Instruments Have Changed Hands* (Leiden: Brill, 2016), 1–20, at p. 5.

3 Jim Bennett, "Shopping for instruments in Paris and London," in Pamela Smith and Paula Findlen, eds., *Merchants and Marvels: Commerce, Science, and Art in Early Modern Europe* (London; New York: Routledge, 2002), 370–395, at p. 372.

4 Henri Lefebvre, *The Production of Space*, Donald Nicholson-Smith, trans. (Oxford: Wiley-Blackwell, 1991).

5 Peter Arnade, Martha Howell and Walter Simons, "Fertile Spaces: The Productivity of Urban Space in Northern Europe," *The Journal of Interdisciplinary History* 32 (2002): 515–548, at p. 541.

households, and employed in their businesses, reveal significant details about their working lives and spaces. This allows us to recapture something of the artisanal experiences of those who existed on the margins of the trade; individuals like mathematical instrument maker Samuel Bellinger, who testified in the 1730s that: 'I am a Mathematical Instrument-Maker, and no House-keeper; I live with my Mother.'[6] Existing studies of instrument makers focus almost exclusively upon established and highly successful producers, those with the most elite contracts, and clientele. This chapter attempts a broader social scope.

1   Organisation of Space

Testimony from court cases, especially narratives concerning robberies of shops, and also rare inventories, together shed light upon the basic organisation of spaces inhabited by instrument makers. What emerges is a varied picture of working environments and experiences. A fluid boundary between domestic, working, and commercial spaces was the common experience for most metropolitan instrument makers across the long eighteenth century.[7] Households were highly complex in London, often containing – in addition to the householder and his family – servants, apprentices and lodgers.[8] Typically, there was no clear spatial demarcation between the working and domestic lives and activities of householders. From the second half of the eighteenth century, the most exclusive instrument makers, normally involved in the design and creation of large apparatus, did enact a gradual separation of domestic, working, and trade activities. This experience at the top end of the market was however atypical. We begin here with the spatial layout and character of the homes and working environments of the vast majority of makers and sellers of instruments, before moving onto an examination of the exclusive minority who created a spatial distinction between making and selling.

As was customary, those compiling the late seventeenth-century inventory of clockmaker Thomas Wise made no spatial distinction between his working

---

6   Trial of John Birt, January 1728, *Old Bailey Proceedings Online* (www.oldbaileyonline.org, version 7.2, 28 January 2018), t17380113–12.

7   Peter Guillery, *The Small House in Eighteenth-Century London* (New Haven, CT: Yale University Press, 2004), 66.

8   Amanda Vickery, "An Englishman's Home is his Castle? Thresholds, Boundaries and Privacies in the Eighteenth-Century London House," *Past and Present* 199 (2008): 147–73, at p. 150. It is likely though that multiple occupancy was a stronger trend in west and central London, than in south and east eighteenth-century London. See Guillery, *The Small House in Eighteenth-Century London*, 34–35.

and commercial sites; they listed all the goods found "in the shop and worke shop."[9] At the Old Bailey, watch maker John Lambert spoke of how "in the evening, the prisoner and another person came together into my shop [...] the prisoner sat down upon the seat where I generally work."[10] A court case of 1770 reveals that mathematical instrument maker John Mollison in Blackfriars had, within the same building, domestic and working spaces: domestic activities were located towards the rear of his house, commercial activities at the street side. This had been the typical organisation of artisanal and trade spaces since the late-medieval period. Mollison explained to the court that, "My wife was alarmed by the noise of breaking the beaufet [cupboard or sideboard] between one and two in the night, in the lower parlour fronting the street, close by the window: we lay, in the back parlour." Following the robbery, and with the thief having fled from the scene, Mollison

> found my bureau and book-case open, which were shut over night, but not locked; all the drawers were taken out: they had taken away a silver watch, capped, and jewel, a coral, set in silver, two guineas, a pair of knee-buckles.[11]

It is not unlikely that Mollison's 'lower parlour' was used for entertaining more privileged consumers, as was common practice in eighteenth-century commercial/domestic sites. Typical too was "storing 'shop goods' in rooms throughout the house."[12] The inventory of clock maker Thomas Taylor shows that this artisan kept workshop tools in his kitchen, and raw materials in the cellar.[13]

A grand larceny case tried at the Old Bailey in September 1768 concerning the theft of property belonging to watch maker William More, is highly revealing of the multifunctional nature of sites of instrument making and selling. In his statement to the court More explained that "I live in Moorfields, and am a watch maker; we keep two different shops, I a watch-maker's shop and my wife a chandler's shop." Further testimony from More and his daughter-in-law,

---

9   Wise, Thomas, Citizen and Clockmaker, 1694–1713, Court of Orphans, London Metropolitan Archive (LMA), CLA/002/02/01/2246.

10   Trial of John Bailey, September 1780, *Old Bailey Proceedings Online* (www.oldbaileyonline.org, version 7.2, 28 January 2018), t17800913–3.

11   Trial of John Underwood, William Wharton, May 1770, *Old Bailey Proceedings Online* (www.oldbaileyonline.org, version 7.2, 28 January 2018), t17700530–23.

12   Jon Stobart, Andrew Hann, and Victoria Morgan, *Spaces of Consumption: Leisure and Shopping in the English Town, c. 1680–1830* (London: Routledge, 2007), 117.

13   Taylor, Thomas, Citizen and Clockmaker, 1678–1693. Court of Orphans, LMA, CLA/002/02/01/2127.

Sarah Cope, established the spatial organisation of his home and work spaces in more depth. Situated on "the corner of Crown-alley", there was one street entrance to both shops, and a kitchen lay between the two commercial sites. Cope explained that, "the [watchmaker's] shop lies on one side of the kitchen, we go through the kitchen into the shop." Later she elaborated, "no body can go out of the chandler's shop to the shop where the watches are kept without coming through the kitchen."[14] Multi-trade households were not at all unusual.[15] As prosecutors and witnesses in court, men and women often spoke of spaces from which multiple trades were conducted: one victim of burglary stated, "I am a mathematical instrument maker, and my wife keeps a hoiser and haberdasher's shop."[16] Joseph Ihon of East Smithfield was both a publican and a watch-maker, practising both trades within his "dwelling house" on Nightingale Lane. Describing the robbery of his property in the summer of 1789, Ihon explained that on the ground floor of the house in "a back room" he had a bar, and "up one pair of stairs, in a small room adjoin to a sleeping room" he constructed watches. Pretending to be a customer at his bar, a thief purportedly crept up the stairs of Ihon's property and stole watches "hanging on hooks, on a work board" in the small first storey room.[17]

The sources from London's principal criminal court suggest that towards the end of the eighteenth century, for the most successful and affluent instrument makers, there was the development of a much clearer demarcation of space, and separation of varied working practices. Benjamin Messer's description of the robbery of his property in Bell-dock, Wapping in the 1780s, reveals a deliberate separation between 'work' and 'sale' spaces.[18] Messer told the court of how "on 17th November last, about half past eight, the maid rang the bell from the sale-shop to the work-shop; before I went down to see if any thing was wanted, the bell rang again."[19] From the relative isolation of the work-shop Messer was unaware of the events unfolding in his distinct commercial

14   Trial of John Farrow, September 1768, *Old Bailey Proceedings Online* (www.oldbaileyonline
      .org, version 7.2, 28 January 2018), t17680907–20.
15   Amanda Flather, "Space, Place, and Gender: The Sexual and Spatial Division of Labour in
      the Early Modern Household," *History and Theory* 52 (2013): 344–360, at p. 358–59.
16   Trial of James Mann, December 1789, *Old Bailey Proceedings Online* (www.oldbaileyonline
      .org, version 7.2, 28 January 2018), t17891209–27.
17   Trial of Nicholas Rogers, October 1789, *Old Bailey Proceedings Online* (www.oldbaileyonline
      .org, version 7.2, 28 January 2018), t17891028–1.
18   This was a trend across eighteenth-century luxury trades. See Kathryn Morrison, *English
      Shops and Shopping: An Architectural History* (London: Paul Mellon Centre for Studies in
      British Art, 2003), 36.
19   Trial of Samuel Harding, William Archer, December 1787, *Old Bailey Proceedings Online*
      (www.oldbaileyonline.org, version 7.2, 28 January 2018), t17871212–87.

space. The internal structure of Messer's building, and the concomitant separation between making and selling, must have had the effect of divorcing the mechanical (and potentially noisy and noxious) processes of making, from the rather more polite and refined experience of shopping.[20] Like the exclusive instrument makers discussed by Jim Bennett in his analysis of seventeenth- and eighteenth-century travellers' accounts of London, these artisans had the unusual luxury of "A shop [as ...] an intermediate space between the street and the workshop."[21]

The case of the burglary of John and Edward Troughton's property (by a small group of men) in 1802 is clear evidence that these immensely successful instrument makers had fully separated their domestic site and activities, from workshops and commercial spaces. This spatial layout was evidently a response to an increased demand from government, institutional, and independent customers, and a reflection of the sheer size of the projects undertaken, and instruments assembled. Edward Troughton told the court that at "No. 136, Fleet-street; my brother John and I sleep and live there, but we have another house in Peterborough-court [adjacent to the aforementioned property], which is only for work-shops and ware-rooms, and an errand boy only sleeps there for the purpose of letting the men in."[22] It is notable that the Troughton's domestic space, No. 136 Fleet Street, which they had inhabited since 1782, had previously been occupied by both Benjamin Coles (father and son), and before them, Thomas Wright, all renowned and highly successful instrument makers.[23] This succession of instrument makers at the top of their profession occupying the same premises must have been well-known to contemporary institutional and private customers, and offers the distinct possibility that not just certain neighbourhoods (the West End), or streets (Fleet Street), but particular commercial spaces acquired symbolic notoriety for being sites of skill and innovation in the instrument making trade.

What are we to make of the evidence of the spatial organisation of artisan dwellings presented so far? First, physical separation between the activities of instrument manufacture, general domestic duties, and commercial practices,

---

20   This was a common trend within luxury shops, see Claire Walsh, "Shop Design and the Display of Goods in Eighteenth-Century London," *Journal of Design History* 8 (1995), 157–76, at p. 160.

21   Bennett, "Shopping for Instruments in Paris and London," 388.

22   Trial of William Bean, February 1802, *Old Bailey Proceedings Online* (www.oldbaileyonline.org, version 7.2, 28 January 2018), t18020217–70.

23   Anita McConnell, "From Craft Workshop to Big Business – The London Scientific Instrument Trade's Response to Increasing Demand, 1750–1820," *The London Journal* 19 (1994): 36–53, at p. 41.

was a rarity. Second, since instrument-makers typically inhabited un-separated spatial sites of work and domestic and family life, observation and knowledge of a broad range of instruments, and demonstrations of such things, must have been everyday sights in early modern households across the metropolis. Thus, it was not simply as display pieces in the homes of England's "better sort" and gentry that "scientific instruments" were visible and accessible. Third, the range of spatial practices undertaken within a single household suggests that some rooms in instrument-making premises must have taken on a different character – for example, as a working space, a shop, a site of instrumental demonstration, food preparation, or socialising – depending upon the time of day, the persons present, and the activities undertaken.[24]

2   Workshops and Working Practices

It is a frustration for historians interested in all artisanal practices that few sources allow us to uncover the manufacturing process. This lacuna is largely a reflection of the embodied and tacit nature of workshop learning and innovation, and the craft culture of the 'mystery', the collective secrets of the trade. Certainly, by the late seventeenth century, there was also the added complication of subcontracting, which enabled flexible access to specialist skills, and involved very complex networks of manufacture which are hard to map or trace with any accuracy.[25] But unusually crime records do shed light upon working conditions and relationships, within, and between, urban workshops. In describing the circumstances in which products and tools were stolen from his workshop in 1779, for example, Jesse Ramsden outlined the working practices in his workshop in Piccadilly, and the sheer scale of his business operation: "I am a mathematical and optical instrument-maker; I employ a great many workmen, and each workman has his own private drawer where he keeps his tools locked up."[26] Ramsden's large workforce was necessary for undertaking the design and construction of observatory and surveying apparatus and, as Anita McConnell has shown, at least thirty-five men were at some stage

---

24   A parallel argument about the flexibility of space in relation to eighteenth-century homes and leisure spaces is made in Benjamin Heller, "Leisure and the Use of Domestic Space in Georgian London," *The Historical Journal* 53 (2010): 623–45, at p. 628.

25   Giorgio Riello, "Strategies and Boundaries: Subcontracting and the London Trades in the Long Eighteenth Century," *Enterprise and Society* 9 (2008): 243–280.

26   Trial of Peter Kelly, January 1779, *Old Bailey Proceedings Online* (www.oldbaileyonline.org, version 7.2, 28 January 2018), t17790113–15.

employed by this maker.[27] From the Old Bailey sources three key themes relating to working practices emerge. First, the high-turnover of workers in large and medium-sized instrument-making establishments; second, the processes and networks through which subcontracting was carried out, and third, the working experiences of those engaged in subcontracting work.

The Old Bailey records give us an indication of the high turn-over of workers within the shops of established makers, and by implication the diffusion of often highly specialized skills throughout the metropolis. Passing comments made by instrument makers about their employees shed light upon craftsmen whose economic survival and social capital was rather more precarious than their own. Edward Troughton spoke of one William Bean who "had been in my service five or six months, and had quitted it about a month or five weeks; he knew the house near as well as I did." It was Bean's intimate knowledge of the working and commercial space which facilitated his burglary of Troughton's shop.[28] Jesse Ramsden mentioned a man who "had been with me this last time, about six months, as a foreman. He lived with me before, about a year and a half." "Philosophical Instrument Maker" David Barclay spoke of a man in his workshop who had "been out of Business; on the 11th May he work'd for me [...] he [then] went away from me [...] He has work'd with me since; the 27th of October was the last Time."[29] Speaking of an alleged thief, John Brasset, watchmaker, claimed that "this lad had three masters before he came to me, I employ many journeymen and apprentices, and can hardly keep my payments."[30] Here we get glimpses of individuals who moved in and out of employment in the premises of more established artisan-retailers.

Turning to the processes and networks of subcontracting, and describing the prosecution of an alleged theft of materials from his workshop, mathematical and optical instrument maker Edward Nairne outlined the method of subcontracting through which brass, the key material for instruments, came to his workshop at Cornhill near the Royal Exchange.[31] Nairne produced a book in court, and explained that the volume "is carried to the founder; and when he sends a parcel of cast brass, he enters it in this book." Evidently the casting

27   Anita McConnell, *Jesse Ramsden (1735–1800): London's Leading Scientific Instrument Maker* (Aldershot: Ashgate, 2007), Ch. 4.

28   Trial of William Bean, February 1802, *Old Bailey Proceedings Online* (www.oldbaileyonline .org, version 7.2, 28 January 2018), t18020217–70.

29   Trial of Abraham Davenport, December 1737, *Old Bailey Proceedings Online* (www .oldbaileyonline.org, version 7.2, 28 January 2018), t17371207–24.

30   Trial of Douglas Wyre, May 1759, *Old Bailey Proceedings Online* (www.oldbaileyonline.org, version 7.2, 28 January 2018), t17590530–5.

31   The surname has been transcribed as 'Navine' but this is almost certainly an error.

of instrument parts was subcontracted, and Nairne's particular patterns were kept by the founder. In other cases heard at the Old Bailey, founders also recognised the work of different masters by sight in court. So far, this process seems straightforward enough; but Nairne was suspicious that the parcel of brass from the founder was underweight, and searching the room of journeyman David Macauly "I found one of the pieces cast from my pattern."[32] Watchmaking in particular was a trade dispersed into multiple parts, skills, and techniques. John Perkins, living on Snow-Hill, stated that "I am a watch-maker, and sell the various materials that compose a watch, and tools that watch-makers use." Later in his testimony Perkins tellingly admitted that he could not swear that the object presented in court came from his own shop because "the movement goes through so many alterations in finishing."[33] This speaks to the multiple hands and thus spaces engaged in the production of devices and instruments.

Crime records occasionally reveal the life experiences of those engaged in subcontracting; often at the margins of the commercial instrument market. These men lacked the status of a retail space of their own at a good London address. Renting a single room in a house, without his own household or business, James Stansbury appears to be just such a craftsman. Tried and found guilty of violent theft, in the course of questioning, Stansbury asserted that "I am a clock-maker', I lodged there [a house in Whitechapel] and work'd in the garret." Despite what appears to be a relatively marginal socioeconomic position, Stansbury firmly asserted his professional identity (and expertise). Stansbury stressed that "My father was a clock-maker, and he left me his tools." Watch-maker William Walpole was also called into court to support the veracity of Stansbury's professional status, testifying "the Prisoner is a clock-maker: he used to work for me when I had clock-work to do, and I have work'd for him in the watch way." And, "I would trust him with any thing in the way of business. I have trusted him with a spring-clock, or table-clock, of 14, 15, or 20 L value. He is a very good hand at spring work."[34] Here we have two workers engaged in a reciprocal exchange of skills across a network of domestic/working spaces. And despite the absence of a shop, independent access to consumers, or prominent family connections, these makers were assertive in their claims to expertise.

---

32   Trial of David Macauly, June 1758, *Old Bailey Proceedings Online* (www.oldbaileyonline .org, version 7.2, 28 January 2018), t17580628–19.

33   Trial of Joseph Langham, Elias Moring, September 1765, *Old Bailey Proceedings Online* (www.oldbaileyonline.org, version 7.2, 28 January 2018), t17650918–39.

34   Trial of James Stansbury, February 1745, *Old Bailey Proceedings Online* (www.oldbaileyonline .org, version 7.2, 28 January 2018), t17450227–12.

3   The Instrument Maker's Shop and Display Strategies

Occasionally shop owners revealed, through their testimony as alleged victims of crime, the material fixtures of their commercial spaces and associated techniques of display. These glimpses into shop layout and design enhance our understanding of the trade of instrument making, including the manufacture, retail, and consumption of instruments. Except for the understanding that the shops of instrument makers probably combined "on the shelf" instruments, ready to buy, and more exclusive and expensive customised objects, made to order, our knowledge of product display is hitherto limited.[35]

Research on the architectural design and decorative elements of seventeenth-and eighteenth-century London and provincial shops has demonstrated the significance of features or material fabrics such as glass windows, plaster mouldings, and gilded cornices, in augmenting and advertising the social status of consumer, and retailer.[36] More particularly, histories of shopping for textiles and luxury products, such as porcelain, argue for the importance of interior display in marketing wares, articulating the professionalism of the seller, and cultivating an atmosphere of polite sociability.[37] Work on goldsmiths and apothecaries has separately shown how distinctive shop designs and display strategies could demonstrate the knowledge and expertise of the retailer, and the authentic nature of their products.[38] What can we say about the shops of instrument makers?

In his testimony concerning a stolen sextant, made of ebony, brass and ivory, valued at 50s, mathematical instrument maker Robert Gota of Wapping described the display of wares inside his shop:

> While I was at breakfast, a young lad knocked at the door and asked me if I had lost any thing, I looked round and saw a glass case door open, and the sextant in it taken out; this glass case stood in the shop, about nine feet from the door, against the shop wall.[39]

Nicholas Meredith, mathematical instrument maker in New Bond Street described a "case of instruments [...] on the counter", which a neighbour later

---

35   Bennett, "Shopping for Instruments."
36   Stobart et al., *Spaces of Consumption*, 112, 126.
37   Walsh, "Shop Design and the Display of Goods."
38   Patrick Wallis, "Consumption, Retailing, and Medicine in Early Modern London," *Economic History Review* 61 (2008): 26–53; Walsh, "Shop Design and the Display of Goods."
39   Trial of Richard Russel, October 1792, *Old Bailey Proceedings Online* (www.oldbaileyonline.org, version 7.2, 28 January 2018), t17921031–59.

saw in the street, in the hand of the suspect. Samuel Cooley, aged nine years, was found guilty of stealing a black shagreen case, brass compasses, a box wooden sector, a box wooden scale, a brass protractor, a steel drawing pen, and three brass compass pieces.[40] Optical instrument maker Philip Brock of Church-row, Aldgate, suffered a considerable loss of stock in September 1818, including a mahogany case, one body of a microscope, four glasses, four ivory sliders and objects, a concave and a plane mirror, and one side illuminator. Brock was

> at work at the window-the articles stated in the indictment were in a case behind me. The prisoner came in to buy a glass for a show. I said I had nobody to serve him [...] he went round on my left hand.[41]

Benjamin Messer's wife explained to the court that the reading glass stolen from her husband's shop "was exposed in the window for sale, before the robbery was committed."[42] Edward Troughton also revealed something about the interior organisation and display of his ware-rooms, stating "in the shop there is but one drawer kept locked, in which I generally keep small valuable items, and which had been wrenched from the bench."[43]

The impression we get from this collective testimony is of a wide range of products for sale in individual establishments, and varied of techniques of display. Instruments were shown-off behind glass windows, exposed within glass cases, stored within wooden cases on shop counters, and locked away within drawers. Items in eighteenth-century shop windows "were vital in attracting passers-by" and giving potential consumers "a taste of what lay inside the shop."[44] Watch makers in particular testify to hanging their instruments on hooks behind window glass. Maker Thomas Sutherland had an instrument stolen "by a violent blow made against the window, it hung up by small brass hook, within an inch of the window."[45] Likewise, William and Thomas Collett had "a

---

40   Trial of Samuel Cooley, May 1778, *Old Bailey Proceedings Online* (www.oldbaileyonline .org, version 7.2, 28 January 2018), t17880507–9.

41   Trial of David Lazarus, October 1818, *Old Bailey Proceedings Online* (www.oldbaileyonline .org, version 7.2, 28 January 2018), t18181028–90.

42   Presumably the display of this object in 'the window' of the shop, as opposed to within a glass case, related to the (lesser) value of the object. Messer himself referred to the merchandise stolen as 'trifling things'. (Trial of Samuel Harding, William Archer, December 1787, *Old Bailey Proceedings Online* (www.oldbaileyonline.org, version 7.2), t17871212–87.)

43   Trial of David Macauly, June 1758, *Old Bailey Proceedings Online* (www.oldbaileyonline .org, version 7.2, 28 January 2018), t17580628–19.

44   Stobart et al., *Spaces of Consumption*, 116.

45   Trial of Thomas Colbrook, February 1786, *Old Bailey Proceedings Online* (www .oldbaileyonline.org, version 7.2, 28 January 2018), t17860222–5.

great many watches then at the window."[46] Instrument maker John Marshall spoke of products "in the inside of the shop, beside my work-board, within six or seven inches of the show-glasses. It is a kind of bow window."[47]

Shopping in the eighteenth century was a negotiation between customer and retailer, and thus the (legitimate) opening up of glass cases and unlocking of drawers must have depended upon the impression made by the consumer upon the retailer. What is striking from this testimony is how few instruments were in fact locked away, out of the reach of consumers, or opportunistic thieves. Even expensive (and portable) instruments were openly on show in shops. One man nearly succeeded, for example, in carrying a telescope out of Thomas Parnell's shop. The thief was only thwarted by the unexpected dimensions of the telescope: the man "tried to take up the telescope, and put it under his clothes, but it was too high above his clothes." On the other hand, noticing that the shop-owner was distracted at his workstation in the window, David Lazarus stole a microscope from a fixed case standing out of the artisan-retailer's sight.[48]

Necessarily absent from sources produced by a judicial institution are more "legitimate" instances of browsing for instruments as consumer objects. Shopping in general required the purchaser to use a variety of senses, including touch, sight, smell, and even taste, to ascertain quality and workmanship.[49] But how much more important sensory interaction must have been for the selection and purchase of instruments, for these were consumer products whose significance and functionality often depended upon a range of applied sensory knowledge. We have the accounts of some of the most elite shoppers in the market – Bennett's genteel and aristocratic foreign visitors to London – but what of more humble and quotidian interactions in the instrument maker's shop by 'middling' and 'lesser' sorts? This is an aspect of the history of instruments which deserves much greater attention, and can only be touched upon here. For now, it is worth observing that in describing interactions which later went badly wrong, instrument makers inadvertently revealed everyday patterns of showing, handling, and examining instruments. An (alleged) customer usually informed the artisan-retailer of his requirements, then he or she was presented with an instrument, or multiple instruments, which fitted

46   Trial of Michael Love, January 1795, *Old Bailey Proceedings Online* (www.oldbaileyonline.org, version 7.2, 28 January 2018), t17950114–9.

47   Trial of William Davidson, Mary Griffin, February 1780, *Old Bailey Proceedings Online* (www.oldbaileyonline.org, version 7.2, 28 January 2018), t17800223–20.

48   Trial of David Lazarus, October 1818, *Old Bailey Proceedings Online* (www.oldbaileyonline.org, version 7.2, 28 January 2018), t18181028–90.

49   Kate Smith, "Sensing Design and Workmanship: The Haptic Skills of Shoppers in Eighteenth-Century London," *Journal of Design History* 25 (2012): 1–10.

the description. It was usual for the buyer to then handle and examine the instrument(s) closely; John Farrow looked at a proffered watch, and to check its working.[50] An extraordinary late eighteenth-century case involving instrument makers in an alleged treason plot including an air gun,[51] contains details of instrumental demonstration within a maker's shop. Mathematical instrument maker David Cuthbert, with premises in Greyhound Court, Arundel Street, off the Strand, showed watchmaker Thomas Upton the workings of an air gun that he had constructed, and explained the properties of air:

> Q. Do you recollect any conversation with him about the properties of air? - A. Yes; he saw an air-pump lying in the shop, I explained it to him as well as I could. I shewed him an air-gun, and I explained that to him [...] Q. Did he handle the gun? - A. He looked at it, and viewed it, and said it was a handsome piece.[52]

Though the nature of the trial was highly unusual, the details of everyday workshop interactions with instruments of experimental philosophy are revealing.

4   Identifying and Asserting Expertise and Authorship

A final theme to be drawn out from the sources relates to languages of expertise employed by those living, working, and selling within these metropolitan spaces. On rare occasions instrument makers were called upon to be, in effect, expert witnesses in Old Bailey trials. A case heard in the 1760s pertaining to coining offences drew upon the workshop knowledge of several optical, mathematical, and clockmakers. Robert Featley, John Hunter, and George Hodgson were cross-examined in court on the use of tools employed in their trade (though in this instance illicitly used to file coins); providing evidence these artisans gave rare insights into the technicalities of their working practices,

---

50   Trial of John Farrow, September 1768, *Old Bailey Proceedings Online* (www.oldbaileyonline .org, version 7.2, 28 January 2018), t17680907–20.

51   This case-study is explored in much greater depth in Jim Bennett, "Wind Gun, Air Gun or Pop Gun: The fortunes of a philosophical instrument," in Lissa Roberts, Simon Shaffer, and Peter Dear, eds., *The Mindful Hand. Inquiry and Invention from the Late Renaissance to Early Industrialisation* (Amsterdam: Koninklijke Nederlandse Academie van Wetenschappen, 2007), 221–45.

52   Trial of Robert Thomas Crossfield, May 1796, *Old Bailey Proceedings Online* (www .oldbaileyonline.org, version 7.2, 28 January 2018), t17960511–1.

and even recognised instrumental innovation. Optical instrument maker Robert Featley examined the tool by sight and touch in the courtroom and declared that "this [tool] is proper for several particulars, microscopes, and things in our way." John Hunter, said to be "conversant in mathematical instruments and clock-work both", was asked "To what purpose is such an instrument as this applicable?" Hunter replied that

> We have several milled nuts, both in the mathematical and clock way; this is more useful than any thing we use, for any thing that will go into it; it is as great an improvement as ever I saw; the edges or jestering nuts for regulating clocks might be done with this.[53]

More commonly, artisanal practitioners – as prosecutors and witnesses – had to draw upon their particular knowledge and expertise to assert that the instruments presented in court (and typically found in the possession of the accused) were the handiwork of their workshop. In the process of identifying stolen things as their own, there was also the opportunity for demonstrating knowledge about instrument designs and materialities. Optician William Arnold asserted that though the glasses produced in court had "no shop-mark", he was absolutely certain that they were his creation: "one of them is a very particular one; the object is let in different to what they generally are, and the eyeglass is larger than it ought to be; I have no doubt of their being mine."[54] A member of Jesse Ramsden's sizeable workforce, Matthew Berge, swore to the court that two sexton glasses which were found in the prisoner's lodging were unquestionably the property of his master: "I know them to be Mr. Ramsden's by the size; they are the size we always make them."[55] Presented with a watch in court, maker Richard Delahoy showed off his knowledge of materials:

> Q. Is that a shagreen outside case?
> Delahoy. No; it is a fish skin.
> Q. If I came to buy a watch, would that pass for a shagreen case?
> Delahoy. Yes.

---

53   Trial of William Guest, September 1767, *Old Bailey Proceedings Online* (www.oldbaileyonline.org, version 7.2, 28 January 2018), t17670909–41.

54   Trial of Jacob Solomons, September 1800, *Old Bailey Proceedings Online* (www.oldbaileyonline.org, version 7.2, 28 January 2018), t18000917–91.

55   Trial of Peter Kelly, January 1779, *Old Bailey Proceedings Online* (www.oldbaileyonline.org, version 7.2, 28 January 2018), t17790113–15.

In a case of 1773 concerning the theft of a watch, worth 30s., watch-maker James Stoddard was brought into court to ascertain the authenticity of the object. Stoddard's involvement is worth quoting in full, as it shows how the trial process might even involve the demonstration of intricate technical and experiential knowledge.

> James Stoddard. I am a watch-maker. (The watch shewn him). I know Alexander Williamson [prosecutor, and owner of the stolen watch]; to the best of my knowledge this is the watch I sold him two years ago; the name seems to have been erased, and another put in; to the best of my judgement it is the watch; if I was indulged to take it to pieces I could certainly tell whether the name has been erased.
>   Q. How long would you be in taking it to pieces?
> Stoddard. Ten minutes in a convenient place. (He takes the watch home and returns with it in a short time).
> Q. Have you examined the watch?
> Stoddard. Yes; I have taken it to pieces; I can swear positively that it is my work, though the name is erased.[56]

Wider members of the artisan-retailer household could also be called upon to identify particular instruments through distinguishing design features, or marks. The wife of mathematical instrument maker Benjamin Messer identified instruments taken from her husband's shop: in 1787 she "proved the property to be her husband's, by a light mark in the tortoise-shell, near the hinge, and that it was exposed in the window before sale, before the robbery was committed."[57] Again in 1794, "I know them by a private mark, which I always mark them with."[58] Certainly subjectivity was involved in this process: Richard Ireland Thorogood, surgical instrument-maker, claimed that "I know the [stolen] spatulas by being the work in my house; the files I know by the private mark."[59] Mathematical and optical instrument maker Edward Nairne identified materials stolen by a former employee, as one of the pieces was said to be "cast from my pattern." Thomas Bardin Globemaker of Salisbury-square, Fleet Street claimed that he had "no doubt" that the items presented in court

---

56   Trial of Thomas Price, September 1773, *Old Bailey Proceedings Online* (www.oldbaileyonline.org, version 7.2, 28 January 2018), t17730908–44.

57   Trial of Samuel Harding, William Archer, December 1787, *Old Bailey Proceedings Online* (www.oldbaileyonline.org, version 7.2, 28 January 2018), t17871212–87.

58   Trial of Daniel Lemon, Elizabeth Newton, April 1794, *Old Bailey Proceedings Online* (www.oldbaileyonline.org, version 7.2, 28 January 2018), t17940430–101.

59   Trial of Henry Wilson, January 1789, *Old Bailey Proceedings Online* (www.oldbaileyonline.org, version 7.2, 28 January 2018), t17890114–66.

belonged to him "because there are few in the trade, and each in the trade have their own patterns, and the person who makes the patterns is here."[60]

Positively identifying objects presented in court was a key element in successfully pursuing a criminal case, but for instrument-makers in particular, the question of authorship was especially significant against the backdrop of widespread subcontracting, and the increasing use of patents as a means of protecting authorship and commercial advantage.[61] Identifying the particular maker, or even workshop from which an instrument originated could be challenging. Moreover, the signature on a purchased instrument might well refer to the vendor, not a manufacturer. As Crawforth has argued, "These false impressions became so widespread that from the late eighteenth century onwards several makers felt it necessary to claim, on their trade cards, that they were not just retailers."[62] The evidence from Old Bailey trials shows that there was a further complication to this instrument/authorship issue: the re-sale of second-hand (and frequently stolen) objects was clearly facilitated by erasing the marks of the genuine maker. In a case of 1745 a horizontal sundial made by Thomas Wright, who traded at the sign of the Orrery and Globe, was stolen from a gentleman's garden. In this instance the thief attempted to erase all identifying marks on the dial, but while he succeeded in eradicating the arms of the sundial's owner, he failed to erase the maker's name, because it was "engraved so deep."[63] In the late seventeenth century, the practice of erasing the genuine maker's name from an instrument and inscribing it with a new or fabricated maker's name was an issue which was frequently raised in the courts of livery companies too. During the 1670s and 80s the Clockmakers' Company repeatedly attempted to mediate in trade disputes concerning "false" and "invented" names on watches and clocks.[64]

5   Conclusion

Criminal sources present us with a unique window upon the spaces of London's scientific instrument trade. Court room testimony conjures up vivid impressions of the physical layout of homes, workshops, and businesses, the

60   Trial of George Lane, September 1805, *Old Bailey Proceedings Online* (www.oldbaileyonline.org, version 8.0, 03 February 2021), t18050918–22.

61   Mario Biagioli, "From Prints to Patents: Living on Instruments in Early Modern Europe," *History of Science* 44 (2006): 139–189.

62   Crawforth, "Evidence from Trade Cards for the Scientific Industry," 478.

63   Trial of William Carter, March 1745, *Old Bailey Proceedings Online* (www.oldbaileyonline.org, version 7.2, 28 January 2018), t17450530–2.

64   Guildhall Library, London, MSS 2710/1, fol. 277; 2710/2, fol. 27v.

multifunctional nature of these sites, and the material arrangement of these dynamic spaces. Reading the primary evidence against the grain also reveals a rich language of expertise and ownership employed by struggling journeymen and established masters alike.

Only artisan-retailers at the very top of the profession, like the Troughton brothers, were able to physically demarcate domestic, working, and commercial spaces. This separation of the practices of manufacture and trade might have had the effect of elevating elite makers from the realms of the "mechanical" to "polite" society. However, since witnessing the manufacture of instruments was desirable, in itself, for certain consumers with an interest in natural philosophy, this will have to remain a tentative conclusion, and a theme which deserves further attention. More detailed research might also be undertaken into the interactive nature of shopping and browsing for scientific instruments in relatively humble working and commercial spaces.

The overall impression from the Old Bailey sources is of a great variety a highly specialized skills, dispersed between densely interconnected domestic, working, and commercial spaces. Knowledge and expertise concerning instrument making in the metropolis was not simply located in the large proto-factories, on which so much ink has been spilt, but could also be found in humble parlours and garrets. Moreover, lacking the usual features of artisanal status - such as a prominent position in a guild, or a prestigious trading address - did not mean that craftsmen were inhibited from asserting expertise. The historiographical focus upon a handful of renowned and unusually affluent eighteenth-century instrument makers has hitherto obscured the broader social and spatial dynamics of the trade.

CHAPTER 5

'My Collection in All Its Branches': The Imagined Space of Early Modern Scientific Correspondence

Alice Marples

In 1729, the physician and antiquary Dr. Cromwell Mortimer moved from Hanover Square to Bloomsbury at the request of his friend, Sir Hans Sloane. On the 23 July of that year, Mortimer had apologised in a letter to Dr. Edmund Waller, a senior fellow at St. John's College, Cambridge, for not having yet shown the Royal Society the "curious leaden bone" that Waller had sent a few months previous. Aside from trying (and failing) to make replicas from the bone, Mortimer had been preoccupied with his new living arrangements:

> I have the pleasure of being at Sir Hans at all leisure hours of the day, continually entertained with new curiosities in his prodigious collection, and having the opportunity of the use of his library, as well as his ingenious and learned conversations.[1]

Among many other shared projects, the two physicians spent three months in 1733 studying a female beaver that Sloane kept in his garden, making observations about its behaviour in captivity compared with what they knew about its life in the wild. When a dog tragically killed it, they took the opportunity to dissect and draw the animal instead, and Mortimer's account of "Our Beaver" was later published in the *Philosophical Transactions*.[2] Though a full drawing of the beaver has not been found, its specimen was preserved in Sloane's collection,

---

1 John Nichols, *Literary Anecdotes of the Eighteenth Century,* vol. V (London: 1812), 425. Waller's original letter regarding the skeleton had been read to the Royal Society on 5 February 1729 (Royal Society, London: RSBO/14/98).

2 Cromwell Mortimer, "The Anatomy of a Female Beaver, and an Account of Castor Found in Her," *Philosophical Transactions* 38 (1733): 172–183. See also F. W. Gibbs, "Cromwell Mortimer, F.R.S.: Secretary, Royal Society, 1730–1752," *Notes and Records of the Royal Society of London* 7 (1950): 260 [n.b. Gibbs claims that the animal dissected was a badger – I am indebted to Felicity Roberts for providing all references to the beaver].

as listed as two entries in one of his catalogues: "1428. The case of a beaver I kept alive in my garden for some time" and "1429. The inward parts in spirits."[3]

This account of Sloane and Mortimer's shared scientific activities demonstrates the many ways in which early modern collections functioned as physical spaces of research for scholars, allowing them to observe, handle and experiment with objects they might otherwise have no chance of working with or even, in some cases, seeing. They were spaces which allowed for the profitable connection of many different strands of material and intellectual enquiry, bringing together all manner of people, objects, ideas and motivations in their amassment and use. As Maria Zytaruk recently stated, "it was the capacity of the cabinet to accommodate divergent readings of nature and to embody rival systems of knowledge, which makes it a crucial site for the history of early modern science."[4] This understanding of collections has drawn much recent scholarly attention to the development of early modern science and society, driven by increasingly global approaches to intersecting realms of trade and empire, colonial encounter, and knowledge-making.[5] Related turns towards the visual and material in the history of science and medicine have been valuable in emphasising the many different lives an object could have, and how these were translated and transformed across different geographical, social, experiential and material spaces.[6] Increasingly, too, this approach has extended to the exploration of entire collections themselves, their changing composition and uses across time and space, as well as their contemporary

---

3   Natural History Museum, Sloane Catalogue, Fossils, vol. V, f.379.

4   Maria Zytaruk, "Cabinets of Curiosities and the Organization of Knowledge," *University of Toronto Quarterly* 80 (2011): 1–23, at p. 3.

5   David Wade Chambers and Richard Gillespie, "Locality in the History of Science: Colonial Science, Technoscience and Indigenous Knowledge," *Osiris* 15 (2000): 221–240; Geri Augusto, "Knowledge Free and 'Unfree': Epistemic Tensions in Plant Knowledge at the Cape in the Seventeenth and Eighteenth Centuries," *International Journal of African Renaissance Studies* 2 (2007): 136–182; Pratik Chakrabarti, *Materials and Medicine: Trade, Conquest and Therapeutics in the Eighteenth Century* (Manchester: Manchester University Press, 2010); Sujit Sivasundaram, "Sciences and the Global: On Methods, Questions, and Theory," *Isis* 101 (2010), 146–158; Kathleen Murphy, "James Petiver's 'Kind Friends' and 'Curious Persons' in the Atlantic World: Commerce, Colonialism and Collecting," *Notes and Records of the Royal Society* 74 (2019): 259–74.

6   Pamela H. Smith, Amy R.W. Meyers and Harold J. Cook (eds.), *Ways of Making and Knowing: The Material Culture of Empirical Knowledge* (Ann Arbor: University of Michigan Press, 2014); Anita Guerrini, "The Material Turn in the History of Life Science," *Literature Compass* 13 (2016): 469–480; Anna Marie Roos, "Object Biographies and Interdisciplinarity," *Notes and Records of the Royal Society* 73 (2019): 279–283.

legacies.[7] But in the majority of this work, the focus has generally remained on collections of objects as physical entities: how objects entered the defined space of the collection, through which manoeuvres and for what purposes, how this space was demarcated and maintained, and how valuable collections enhanced capital within wider society.[8] To wit, early scientific collections have been approached almost exclusively as part of a broad conception of the European "centre of calculation" or "house of experiment", with objects removed first from indigenous settings and then from quotidian society, to be installed in scholarly institutions or private houses with restricted access, all the better for exerting power over the knowledge they contained and maintaining both social and scholarly authority.[9]

In this chapter, I seek to break down the physicality of these approaches by demonstrating the importance of correspondence networks as an imagined extension of the early modern scientific collection. In recent years, we have gained a crucial understanding of just how much commerce and natural philosophy alike was shaped by "intelligencers" or "information brokers" who actively maintained vast correspondence networks and gathered all forms of knowledge objects. The huge archives of those who left a paper trail, from humanist polymaths like Marin Mersenne and Henry Oldenburg to committed imperial expanders such as René Antoine Ferchault de Réaumur

---

7   Jim Bennett, "Museums and the History of Science: Practitioner's Postscript," *Isis* 96 (2005): 602–608; Ad Maas, "The Storyteller and the Altar. Museum Boerhaave and its Objects," in Susanne Lehmann-Brauns, Chistian Sichau and Helmuth Trischler, eds., *The Exhibition as Product and Generator of Scholarship* (Berlin: Max-Planck-Institut, 2010); Nicholas Jardine, "Reflections on the Preservation of Recent Scientific Heritage in Dispersed University Collections," *Studies in History and Philosophy of Science* 44 (2013): 737; Boris Jardine, Emma Kowal and Jenny Bangham, "How Collections End: Objects, Meaning and Loss in Laboratories and Museums," *BJHS Themes: How Collections End* 4 (2019): 1–27; Alice Marples, "Scholarship, Skill and Community: Collections and the Creation of 'Provincial' Medical Education in Manchester, 1750–1850," *Journal of the History of Collections* 33 (2021): 505–516.

8   Ilana Kausman Ben-Amos, *The Culture of Gift-Giving: Informal Support and Gift-Exchange in Early Modern England* (Cambridge: Cambridge University Press, 2008); Craig Ashley Hanson, *The English Virtuoso: Art, Medicine, and Antiquarianism in the Age of Empiricism* (Chicago: University of Chicago Press, 2009); Felicity Heal, *The Power of Gifts: Gift-exchange in Early Modern England* (Oxford: University of Oxford Press, 2014).

9   Steven Shapin, "The House of Experiment in Seventeenth-Century England," *Isis* 79 (1988): 373–404; Michael Hunter, *Establishing the New Science: the Experience of the Early Royal Society* (London: Boydell, 1989); Samuel J.M.M. Alberti, "Objects and the Museum," *Isis* 96 (2005): 559–571; Ken Arnold, *Cabinets for the Curious: Looking Back at Early English Museums* (Aldershot: Routledge, 2008); Simon Chaplin, "John Hunter and the 'Museum Oeconomy', 1750–1800" (unpublished PhD dissertation: King's College London, 2009); Edwin D. Rose, "From the South Seas to Soho Square: Joseph Banks's Library, Collection and Kingdom of Natural History," *Notes and Records of the Royal Society* 73 (2019): 499–526.

and Joseph Banks, reveal a vast array of interests, intentions, materialities, spaces, practices and ideas.[10] They also contain many examples of individuals seeking out and working with a range of private and institutional collections. Such networks accumulated and disseminated knowledge in and around a collection, facilitating almost all interaction with it: in this, they represented a discursive space which linked the physical objects of a collection with the diverse meanings and contexts we now understand to be so integral to early modern knowledge making.[11] It was through the space of the collectors' correspondence networks, and not the physical space of the collection itself, that the conflicting motivations or behaviours of each community of users mostly intersected, working to influence one another as well as the collector, the collection, and the knowledge it helped to create. Together, collections and correspondence acted as a "boundary object", uniting diverse people, objects, motivations and practices, and creating space for what Susan Leigh Star has termed "cooperation without consensus."[12] Though limiting boundaries certainly existed within correspondence networks, they were highly flexible and dependent on context, need and material, and extended far beyond those who could physically enter the collection.[13] This meant that individuals, ideas and materials could enter and interact with a collection through correspondence

---

10   David Philip Miller, "Joseph Banks, Empire, and 'Centres of Calculation' in late Hanover-
      ian London," in David Philip Miller and Peter Hanns Reill, eds., *Visions of Empire: Voyages,
      Botany, and Representations of Nature* (Cambridge: Cambridge University Press, 1996);
      Andrea Rusnock, "Correspondence Networks and the Royal Society, 1700–1750," *The Brit-
      ish Journal for the History of Science* 32 (1999): 155–169; David A. Kronick, "The Commerce
      of Letters: Networks and 'Invisible Colleges' in Seventeenth- and Eighteenth-Century
      Europe," *The Library Quarterly* 71 (2001): 28–43; Natasha Glaisyer, "Networking: Trade and
      Exchange in the Eighteenth-Century British Empire," *Historical Journal* 47 (2004): 451–
      476; Simon Schaffer, "Newton on the Beach: The Information Order of Principia Math-
      ematica," *History of Science* 47 (2009): 243–276; Mary Terrall, *Catching Nature in the Act:
      Réaumur and the Practice of Natural History in the Eighteenth Century* (Chicago: Univer-
      sity of Chicago Press, 2013); Justin Grosslight, "Small Skills, Big Networks: Marin Mersenne
      as Mathematical Intelligencer," *History of Science* 51 (2013): 337–374.
11   Alice Marples and Victoria R.M. Pickering, "Patron's Review: Exploring Cultures of Col-
      lecting in the Early Modern World," *Archives of Natural History* 43 (2016): 1–20.
12   Susan Leigh Star and James R. Griesemer, "Institutional Ecology, Translations and Bound-
      ary Objects: Amateurs and Professionals in Berkeley's Museum of Zoology, 1907–39,"
      Social Studies of Science 19 (1989): 387–420; Bruno Latour, "On actor-network theory: A
      few clarifications," *Soziale Welt* 47 (1996): 369–381; Susan Leigh Star, "This is Not a Bound-
      ary Object: Reflections on the Origin of a Concept," *Science, Technology & Human Values*
      35 (2010): 601–617.
13   Lorraine Daston, "The Ideal and Reality of the Republic of Letters in the Enlightenment,"
      Science in Context 4 (1991): 367–386; Anne Goldgar, *Impolite Learning: Conduct and Com-
      munity in the Republic of Letters, 1680–1750* (New Haven: Yale University Press, 1995); Alice

in ways which they were not able to do physically. Ultimately, the lack of fixed conceptual, social, and spatial boundaries worked to create a neutral and eminently malleable space in which all manner of resources could be evaluated, mediated or mobilised.

This can be seen throughout the correspondence of the important eighteenth-century information broker, Hans Sloane, and his contemporaries. His collection operated as both a neutral and authoritative space for various individuals and communities, simultaneously encouraging discussion, experimentation and integration whilst also providing the wide-ranging means by which 'authoritative' knowledge could be established. Materials from Sloane's extensive collections were requested specifically to test theories, advance experiments, or develop publications, and individuals would offer all sorts of information, ideas and objects in exchange, hoping to connect with whatever related materials he held in his collections or activities he might be privy to through his correspondence networks. So, for example, to repay Sloane for giving him a good recommendation following a critical challenge regarding his medical credentials, one William Maynard sent Sloane stones coughed up by a patient with consumption, which he believed "fit to place among [his] Collection of Raritys",[14] while the physician and naturalist Richard Richardson sent Sloane "the Tumour... and Stone that was drawn out of a Womans Anus", along with "a large Corneouse Excrescence which grew upon the thigh of a Crow", an account of brass instruments recorded in Thomas Hearne's Antoninus and his musings on a fossilised tree whose bark had been used for fuel.[15] As Sloane's reputation grew, so did his collection and so too did the reach of his vital epistolary networks. Furthermore, as this chapter will show, the ways in which Sloane collected and organised his collection (as well as disseminated it through his correspondence networks) demonstrates an acute contemporary awareness of the fact that its value lay in the way it connected and combined many forms of material and cognitive activities, and removed authority from any one source. In this way, the early modern collection became a key component of the conceptualisation of scientific objectivity and practice, something which united as well as demarcated the developing disciplines in the eighteenth

---

Marples, "James Petiver's 'Joynt-Stock': Middling Agency in Urban Collecting Networks," *Notes and Records of the Royal Society of London* 74 (2020): 239–258.

14   British Library [hereinafter BL] Sloane MS 4053, f.289. William Maynard to Hans Sloane (17 October 1734, Wigan). Sloane seemingly stepped in to resolve the suspicions of the original letter: BL Sloane MS 4053, f.257. John Barlow to Hans Sloane (22 August 1734, Manchester).

15   BL Sloane MS 4046, f.79. Richard Richardson to Hans Sloane (Richard Richardson 1 Apr 1721, North Bierley).

century, condensing and transcending the physical boundaries of the natural and human world.[16]

This can perhaps be seen most clearly in highly contested areas of early modern scientific knowledge, such as medicine, where new markets of global *materia medica* and professional service were upsetting established hierarchies of knowledge and practice, and accusations or profiteering or "quackery" were rife at every level of operation. In all the confusion it seemed that no one, learned or otherwise, could possibly claim outright authority over medical knowledge, as the chemist Adrian Huyberts stated when he claimed to have

> lived long enough (almost forty years acquainted with this Art) to see it by improvement in all points turned topsie-turvie, the old Learning belonging to it exploded by Scholars themselves, the old Education in Academies judged incompetent, the places themselves being too narrow to afford much observation or experience, and the manner of life more speculative and notional than Mechanick or laborious, which a Physitian's ought to be.[17]

In such a fraught climate, it was imperative that medical knowledge was collected, tested, evaluated and, where appropriate, adapted. Correspondence was the main way in which this was done. Andreas-Holger Maehle argued some time ago that, rather than a nineteenth-century development, the basic methodology of experimental pharmacology was formed in the eighteenth century through the critical examination of key drugs of the period, such as opium and Peruvian bark, and of important proprietary medicine, such as Mrs Stephens's remedy against bladder stones.[18] The evaluation of remedies was not confined to university medicine and learned scientific societies, but rank and file contributed as well, something which is borne out by the broad-ranging literature on the subject.[19]

---

16   Lorraine Daston, "Objectivity and the Escape from Perspective," *Social Studies of Science* 22 (1992): 597–618; Jennifer Tucker, "Objectivity, Collective Sight, and Scientific Personae," *Victorian Studies* 50 (2008): 648–657; Lorraine Daston, "The Empire of Observation, 1600–1800," in Lorraine Daston and Elizabeth Lunbeck, eds. *Histories of Scientific Observation* (Chicago: University of Chicago Press, 2011).

17   Adrian Huyberts, *A Corner-Stone laid towards the Building of a New Colledge* (London: 1675), 11; Mark Jenner and Patrick Wallis, eds., *Medicine and the Market in England and its Colonies, c.1450-c.1850* (Basingstoke: Palgrave Macmillan, 2007); *Harold J. Cook, Matters of Exchange: Commerce, Medicine, and Science in the Dutch Golden Age* (New Haven: Yale University Press, 2007).

18   Andreas-Holger Maehle, *Drugs on Trial: Experimental Pharmacology and Therapeutic Innovation in the Eighteenth Century* (Amsterdam: Rodopi, 1999), 1–2;

19   Hubert Steinke and Martin Struber, "Medical Correspondence in Early Modern Europe: An Introduction," *Gesnerus* 61 (2004): 139–60; Gianna Pomata, "Sharing Cases: The

Physicians had long been judicious in building up their own stores of medical knowledge, amassing manuscript archives, casebooks, libraries and collections. During Sloane's lifetime, however, the active collection of ephemeral medical material became essential for many physicians and practitioners, and drew a broader range of individuals into the circulation and evaluation of useful medical knowledge. Their collections are testament to this.[20] One example is Sloane MS 3948, described as "a collection of loose papers and letters from the seventeenth and eighteenth centuries", and while it was likely compiled and bound by later archivists rather than Sloane himself, it demonstrates the range of his interests and the kinds of things that medical practitioners were gathering. There are a number of medical recipes mixed in with prescriptions (including one for a Mrs Pepys, spanning March to June 1728), ships records, draft letters by Sloane himself, a tally list of inoculations conducted, the recommendation of a surgeon to a Professorship at the University of Glasgow, and recipes for the cleaning of shells and the preservation of reptiles, insects and flowers. On the back of the recipe entitled "Mrs Bowles receipts for ruptures, 17thc", Sloane has written: "This is a most Material paper upon the Cure of Ruptures in the Hand writing of Mrs Bowles from whom Benton had what he knew relating thereto."[21] Also in his own hand is the following appetising medical recipe:

> Take 50 snails well washd, 12 whites of new laid egs well beaten, 4 nutmegs shred, the rind of one orange shred, 4 good handfuls of ground ivy call'd alehoof shred, putt these into a galleon of new milk & let them

---

Observationes in Early Modern Medicine," *Early Science and Medicine* 15 (2010): 193–236; Elaine Leong, "Collecting Knowledge for the Family: Recipes, Gender and Practical Knowledge in the Early Modern English Household," *Centaurus* 55 (2013): 81–103; Elaine Leong and Alisha Rankin, eds., *Secrets and Knowledge in Medicine and Science, 1500–1800* (London: Routledge, 2016).

20   Gianna Pomata and Nancy Siraisi, eds., *Historia: Empiricism and Erudition in Early Modern Europe* (Cambridge, Mass: MIT, 2005); Nancy Siraisi, *History, Medicine and the Traditions of Renaissance Learning* (Ann Arbor: University of Michigan Press, 2008); Gianna Pomata, "Sharing Cases: The *Observationes* in Early Modern Medicine," *Early Science and Medicine* 15 (2010): 193–236; Volker Hess and J. Andrew Mendelsohn, "Case and Series: Medical Knowledge and Paper Technology, 1600–1900," *History of Science* 48 (2010): 287–314; Lauren Kassell, "Casebooks in Early Modern England: Medicine, Astrology, and Written Records," *Bulletin of the History of Medicine* 88 (2014): 595–625; Hannah Murphy, "Common Places and Private Spaces: Libraries, Record-Keeping and Orders of Information in Sixteenth-Century Medicine," *Past & Present* 230 (2016): 253–268; Alice Marples, "Medical Practitioners as Collectors and Communicators of Natural History in Ireland, 1680–1750," in John Cunningham, ed., *Early Modern Ireland and the World of Medicine: Practitioners, Collectors and Contexts* (Manchester: Manchester University Press, 2019).

21   BL Sloane MS 3984, ff.16–17.

stand over night, then distill them with a gentle fire 'till halfe be drawn of. Drink a quarter pint morning & evening daily sweetened with a spoonefull of balsamick syrup.[22]

Within the same volume, too, there is evidence of heavy revision as to the precise directions for the use of balsamic syrup to soothe the eyes, showing how he returned to correct recipes after trying them and seeing their effects. This continual calibration of knowledge in answer to newly-discovered and often ephemeral information, regardless of composition or creation date, is something we can see in Sloane's published recipe *An Account of a most Efficacious Medicine for Soreness, Weakness, and several other Distempers of the Eyes* (1745).[23]

We can see here how Sloane's professional experience as a physician may have accounted for his perceived willingness to accept anything from anyone, a well-known characteristic which drew much censure and satirical comment from scholars and critics alike.[24] Yet it was precisely the breadth and eclecticism of his collection that encouraged individuals to get in touch with him in the first place, aware that his collection could connect disparate scraps from all forms of enquiry and all ages. One example of this was when a J. Delacoste sent Sloane an excerpt from the second volume of Jean-Baptiste Labat's *Nouveau voyage aux iles de l'Amerique* (Paris, 1722), desiring his opinion. Delacoste had been amazed to discover from it that Amerindians had their own methods to treat disease, and thought this topic was worthy of further exploration, so wrote to Sloane to see if it had ever been treated elsewhere: "If half of what this author affirms to be true, You must know it, if any body."[25] Similarly, Thomas Knight wrote, in 1737, asking if Sloane had any knowledge of anything resembling the uncommon case he had discovered, and begging him to "obviate" any

---

22 BL Sloane MS 3984, f.84.

23 BL Sloane MS 3984, f.92; Arnold Hunt, "Sloane as a Collector of Manuscripts" in Alison Walker, Michael Hunter and Arthur MacGregor, eds., *From Books to Bezoars: Sir Hans Sloane and His Collections* (London: British Library, 2012).

24 Barbara M. Benedict, "Collecting Trouble: Sir Hans Sloane's Literary Reputation in Eighteenth-Century Britain," *Eighteenth-Century Life* 36 (2012):111–142; James Delbourgo, "'Exceeding the Age in Every Thing': Placing Sloane's Objects," *Spontaneous Generations: A Journal for the History and Philosophy of Science* 3 (2009): 44; Miles Ogborn and Victoria Pickering, "The World in a Nicknackatory: Encounters and Exchanges in Hans Sloane's Collection," in Adriana Craciun and Mary Terrall, eds., *Curious Encounters: Voyaging, Collecting, and Making Knowledge in the Long Eighteenth Century* (Toronto: University of Toronto Press, 2019).

25 BL Sloane MS 4046, f.256. J. Delacoste to Hans Sloane (2 July 1722, Bath).

objections to Knight's interpretation of the matter.[26] Sloane's correspondence was seen as a neutral space in which individuals could test their opinions and the results of experiments or initial research against the accumulated knowledge represented by Sloane's collection and broader correspondence networks. As early as 1694, Ray wrote

> I have set down some of my conjectures concerning some of the species of Fishes [described in Sloane's papers], which I offer to your consideration... I would willingly be further assured thereof from yourself, & whether they be known to the Seamen by any names.[27]

Similarly, Samuel Dale continually used Sloane's correspondence and collection to augment new editions of his *Pharmacopoeia* stating, in March 1697,

> I doubt not but divers things relating to the *materia medica* have come to your knowledge since I published that Book... I humbly intreat that favour of you as that if you know where they might be brought you would acquaint Mr Smith of it that he might procure them for me, if not that then you would please to grant me the loan of yours and they shall be thankfully and carefully returned.[28]

Dale wrote to Sloane again in 1733, saying that since "being desirous of making the intended Edition of my Pharmacologia as Servisable unto our Europeans as I can, I have endeavored to find out what ever Species of their *Materia Medica* I had omitted in the former edition," he had been unable to account for many of the things he had discovered and needed Sloane's help once more.[29]

Within this space, individuals could also contribute to broader scholarly discussions which might otherwise be closed to them for reasons of intellectual or social propriety. Only a day after meeting Sloane, for example, one Robert Eaton chose to impart some further knowledge by letter:

> Your mentioning the spitting of blood & ineasiness of lungs & cough struck me with concern & embarrassment what to do, since I knew a

---

26   BL Sloane MS 4043, f.34. Thomas Knight to Hans Sloane (20 February 1737, Caernarfon).

27   BL Sloane MS 4036, f.160. John Ray to Hans Sloane (31 January 1693, Black Notely).

28   BL Sloane MS 4037, f.37. Samuel Dale to Hans Sloane (9 March 1697/8, Braintree).

29   BL Sloane MS 4053, f.52. Samuel Dale to Hans Sloane (19 September 1733, Braintree).

certain & easy method to stop such haemorraging with uttmost conveni-
ence & safety; but for me a Junior & Graduate to mention it to you my
Superior had such a shew of forwardness that in modesty I durst not;
However Gratitude and duty to so Generous a friend oblige me to offer it
you as I thus do in this Vial.[30]

Within the vial which accompanied this letter was a "Balsamick Liquor" which
Eaton swore would stop any internal or external bleeding. He recommended
that Sloane get a Surgeon to experiment with it on a dog, but asked that his
name be concealed throughout any process:

> Sir, make which experiments of it you please, but conceal my name; its
> not a time yet to let it be known; I am too young & too little known; Envy
> & Magisterial Authority of opposite numbers wd crush the reputation of
> so beneficial a medicin. But as I am sure & certain & engage to prove to
> you privatly what I here write I could neither in gratitude nor friendship
> delay putting in your hand a medicine of whose soveraign virtue in your
> & many cases I am assured of.[31]

Through letters, collections could be consulted, wider knowledge could be
imparted, and problematic authorities could be established without damaging
the social or scholarly credit of the individual inquirer, nor any institutional,
intellectual, or commercial body they were attached to. They also provided a
valuable means to obtain professional support and advice away from judge-
mental eyes. Having conducted several further experiments and improvements with his friend Mr Moult, under the guidance of Sir Richard Blackmore,
Eaton later asked for Sloane's advice as to how best to

> dispose of the medicin in a way most honourable both for the Gentle-
> men of the faculty & my self & the good of mankind... Therefore pre-
> suming on your known Prudence & Humanity & as you are now in that
> honourable station as President over the Guardians of health I can't
> doubt of your friendship & advice in so critical a case wherein both the
> success of this medicine & my own reputation in it are so nearly con-
> cerned, having already met with severe discouragement which I should

---

30   BL Sloane MS 4045, f.126. Robert Eaton to Hans Sloane (4 July 1718, Coleman Street).
31   Ibid.

(if it ever becomes necessary) for my own justification be very unwilling to publish.[32]

The projected role of broad collections within the scientific community thus encouraged individuals to write to Sloane before entering the fray of publication or scholarly life. The noted surgeon and natural philosopher Browne Langrish, who first consulted Sloane over a medical case, subsequently sent him a manuscript:

> If you are pleased to approve of it, I am very willing to make some Alterations in it… [reading it in front of the Royal Society would be] my Duty to accept of it, if you think any Theory of Musclar Motion deduced from such Principles as will stand the Test of so learned an audience.[33]

Clearly Sloane did have some reservations, as Langrish followed up:

> Since I had the Favour of your Letter I have thrice repeated the Experiment of tying up the Aorta descendens with all the accuracy imaginable, & find that if the Dog be set down immediately after the aorta is tied up, he can use his lower Parts & walk for a Minute or two, & then a Palsy succeeds.[34]

He went on to explain the differences in his argument to fellow surgeon William Cowper. Sloane clearly urged him to include these in the essay:

> I hope I have deduced my Theory of muscular motion from such Principles as will make it appear rational & consistent; though I am far from thinking it is without Fault, & therefore am willing & ready to correct any of them which my Friends shall advertise me of. – I intend to add a Page or two concerning the Laws of Attraction & Repulsion common to all Matter, whereby the Cause of Elasticity & Contraction of a muscular Fibre may be more fully investigated.[35]

---

32   BL Sloane MS 4045, f.277. Robert Eaton to Hans Sloane (30 December 1719, Coleman Street).

33   BL Sloane MS 4052, f.173. Browne Langrish to Hans Sloane (24 August 1732, Petersfield). For mention of the medical case of Mrs Cole, see: BL Sloane MS 4052, f.301: Browne Langrish to Hans Sloane (16 March 1733, Petersfield).

34   BL Sloane MS 4052, f.180. Browne Langrish to Hans Sloane (12 September 1732, Petersfield).

35   Ibid.; BL Sloane MS 4052, f.184. Browne Langrish to Hans Sloane (17 September 1732, Petersfield).

A few months later, Langrish wrote again:

> The Favours I have already received, and the Encouragement you are always so ready to give to Those who do their best Endeavours towards the Improvement of Natural Knowledge make me presume to send you some Thoughts of mine concerning a new Theory of the Clouds or Vapours; wherein I have shewn that they produce various Effects on the Barometer according as their constituent Particles are in a state of attraction or Repulsion. – I am in Hopes that what I have advanced is deduced from sound Principles in Philosophy, and I don't know but it is intirely new; and if so, I flatter my self it may be acceptable to the illustrious Society I have submitted it to. But if I should be mistaken, I depend, Sir, on your unbounded Goodness to excuse my juvenile Confidence, and to commit what I have wrote to the Flames rather than where I have had the Vanity to direct it.[36]

Langrish eventually honoured Sloane by adding his name to the eventual publication, *Essay on Muscular Motion*, "in Gratitude for the Favours I have received, and as the only Security and Protection against malevolent and ill-natured Critics."[37] Though couched in the traditional forms of flattering patronage, this example demonstrates how Sloane correspondence functioned as sort of filter or informal "peer-review" system for the Royal Society.[38] By getting in touch with Sloane first, individuals could have the benefit of his collections and also be directed in their endeavours. Before embarking on his next project, Langrish wrote again,

> By the last I received a Letter from Dr Hales, who informs me that he has sent my Experiments with the Lauro-Ceresus to you, which I did not design he should have done without my Letter... If you, Learned Sir, approve of my Method, and think it a likely way to discover the Virtues and Properties of other Plants, which we dare not, as yet, use in Physic, I shall readily prosecute the Enquiry, and should be extremely obliged, if you would be pleased to suggest to me what Plants would be fittest for my Purpose, and most likely to reward my Researches.[39]

---

36   BL Sloane MS 4052, f.301. Browne Langrish to Hans Sloane (16 March 1733, Petersfield).

37   BL Sloane MS 4052, f.306. Browne Langrish to Hans Sloane (31 March 1733, Petersfield).

38   Noah Moxham, "Fit for Print: Developing an Institutional Model of Scientific Periodical Publishing in England, 1665-ca.1714," *Notes and Records of the Royal Society* 69 (2015): 241–260.

39   BL Sloane MS 4053, f.135. Browne Langrish to Hans Sloane (13 January 1734, Petersfield).

Through Sloane, Langrish knew that he could be assured his work would be of interest to the Royal Society.

Sloane's authority – his ability to both inform and protect his correspondents – was built in part on the breadth of his collection and the way in which it united so many disparate spaces, authorities and materials, influences and interventions. Access to his collection through correspondence provided the means by which authoritative knowledge could be created, and this was done through the sharing of supposition and strange tales alongside the testing of hypotheses. It was also done through cross-referencing activities which helped to mitigate the various epistemological problems that such a wide-ranging accumulation of miscellaneous sources presented. When most objects and texts entered the collection, they were recorded in Sloane's catalogues: a description of the object (and sometimes additional information regarding source and use) was recorded by Sloane or his assistants alongside a number, labelled on the object itself, that allowed them to find this description and other relevant information in the catalogues. Similarly, Sloane's copies of John Ray's *Historia Plantarum* (London, 1698) and his own *Voyage to Jamaica* (London, 1707) were heavily annotated, cross-referenced with items in his own collection as well as others, and corrected according to new information (some of which was cut and pasted into the printed work itself).[40] James Delbourgo has described the catalogues as "accession registers designed for endless extension to incorporate new items", but they also incorporated new readings of those items, and thus became another form of research.[41] Several scholars have argued that organisational technologies such as journals, lists, notes, boxes and cases both reflected and responded to the diversity of meanings involved in collecting and working with objects, and these "paper technologies" significantly altered the shape of the knowledge produced and circulated through the very structures meant to manage it.[42] As Kim Sloane and Julianne Nyhan have recently demonstrated through their pioneering digital humanities project,

---

40 Edwin D. Rose, "Natural History Collections and the Book: Hans Sloane's A Voyage to Jamaica (1707–1725) and his Jamaican Plants," *Journal of the History of Collections* 30 (2018): 15–33.

41 James Delbourgo, *Collecting the World: The Life and Curiosity of Hans Sloane* (London: Penguin Books, 2017), 259 and 274.

42 Anke te Heesen, "Boxes in Nature," *Studies in History and Philosophy of Science* 31 (2000): 381–403; Staffan Müller-Wille, "Collection and Collation: Theory and Practice of Linnaean Botany," *Studies in the History and Philosophy of Biological and Biomedical Sciences* 38 (2007): 541–562; Staffan Müller-Wille and Isabelle Charmantier, "Natural History and Information Overload: The Case of Linnaeus," *Studies in the History and Philosophy of Biological and Biomedical Sciences* 43 (2012): 4–15.

Sloane's catalogues were more than merely finding tools, and that their intellectual purpose went beyond listing, indexing or valuing, but rather reflected a broader "Enlightenment endeavour to understand the world through objects, text and image… and through the people who made, used and collected these objects throughout history."[43]

Thomas Birch wrote in praise of Sloane's collection after his death:

> In short, the Naturalist will find in this Musaeum almost ever thing, which he can wish, & will be greatly assisted in his Inquiries & Observations by the Catalogue of it in 38 Volumes in fol. & 8 in quarto, containing short Accounts of every particular, with Reference to the Authors, who have treated them.[44]

His comments reveal how collections were increasingly understood as communal resources which could actively facilitate diverse forms of knowledge collection and production across space and through time, and that this capacity was created through correspondence and organisational technologies. Epistemological authority was ultimately placed not only in the physical collection itself (where competing sources sat alongside one another) but also in the archival instruments used to utilise, mediate, and mould its contents, thus commanding the collection's subjective breadth and historical depth.[45] The nature of the collection's continual composition through correspondence networks – both cumulative and collaborative – placed it beyond the control of the individual scholar, beyond even the singular scholarly institution. It became instead a shared resource, and this appreciation can be seen in the efforts made to not only expand collections in the eighteenth century, but also to improve their organisation and records, explicitly to help the circulation of knowledge between various spaces and communities.

Sloane employed people such as Johann Jakob Scheuchzer, Thomas Stack and Cromwell Mortimer to aid in the preservation and organisation of his collection and open its use across the broadest possible spectrum. Mortimer and

---

43   Kim Sloan and Julianne Nyhan, "Enlightenment Architectures: The Reconstruction of Sir Hans Sloane's Cabinet of 'Miscellenies'," *Journal of the History of Collections* 33 (2021): 199–218, at 213.

44   Thomas Birch, "Memoirs Relating to the Life of Hans Sloane," in Alison Walker, Arthur MacGregor and Michael Hunter, eds., *From Books to Bezoars: Sir Hans Sloane and his* Collections (London: The British Library, 2012), 245.

45   Paula Findlen, "The Museum: its Classical Etymology and Renaissance Geneaology," *Journal of the History of Collections* 1 (1989): 59–78; Lorraine Daston, "The Sciences of the Archive," *Osiris* 27 (2012): 156–187.

Stack worked together on Sloane's catalogues of both books and objects from 1729 until the early 1740s. They transcribed a huge number of bibliographic entries and made additions or corrections to existing entries across many catalogues. This collaboration extended beyond the space of Sloane's own library and collection, too: they also worked in this capacity for the Royal Society – in 1739, for example, they were involved a project which involved publishing the Society's registers.[46] It is perhaps no coincidence that the Royal Society underwent a number of vital internal reforms during Sloane's time as Secretary and then President between 1700 and 1740. A committee was organised to re-catalogue and rebuild the Repository, and the *Philosophical Transactions* were revived largely using material from Sloane's own networks, something which encouraged correspondents to once again share their work with the Society and its readership, ensuring its renewed strength in the eighteenth century and the wider encouragement of the circulation of scientific knowledge across British public.[47]

Throughout his life Sloane acquired the collections of other collectors. In 1702 he inherited (in exchange for clearing some debts) the large cabinet of natural and artificial rarities which belonged to the merchant William Courten (sometimes Charleton), who Sloane had met as a young medical student in Montpellier and maintained a friendship with in London.[48] From then on, he seems to have actively sought out the collections of deceased colleagues and contacts. He focused initially on botanical collections. In 1710 he acquired the collection of his long-time botanical antagonist, Leonard Plukenet, and in 1711 he arranged the purchase of German botanist Paul Hermann's herbarium from his wife, Ann, sending the apothecary James Petiver to the continent obtain it. In 1717, Sloane purchased (for a considerable sum of money) the East Indian curiosities, drawing and manuscripts of Engelbert Kaempfer through the royal physician, Dr Johann Georg Steigerthal, and arranged for the Swiss-born naturalist Johann Gaspar Scheuchzer to translate and publish *The History of Japan* in 1727.[49] When Sloane's friend, agent and fellow naturalist, James Petiver died

---

46 Amy Blakeway, "The Library Catalogues of Sir Hans Sloane: Their Authors, Organisation, and Functions," *Electronic British Library Journal* 38 (2011): 1–49; Marie Boas Hall, *The Library and Archive of the Royal Society, 1660–1990* (London: Royal Society, 1992).

47 Alice Marples, "Scientific Administration in the Early Eighteenth Century: Reinterpreting the Royal Society's Repository," *Historical Research* 92 (2019): 183–204.

48 Carol Gibson-Wood, "Classification and Value in the Seventeenth-Century Museum: William Courten's Collection," *Journal of the History of Collections* 9 (1997): 61–77.

49 Basil Gray, "Sloane and the Kaempfer Collection," *The British Museum Quarterly* 18 (1953): 20–23; John Z. Bowers, "Engelbert Kaempfer: Physician, Explorer, Scholar, and Author," *Journal of the History of Medicine and Allied Sciences* 21 (1966): 237–259.

in 1718, his vast and valuable collection was transferred to Sloane wholesale so that he could, he claimed, ensure "that what [Dr. Petiver] hath gather'd together, by a very great and undefatigable Industry, shall not be lost, but preserved and publish'd for the good of the Publick, doing right to his Memory, and my own Reputation."[50] In Cromwell Mortimer's letter to Dr Waller, which opened this chapter, he wrote,

> I must congratulate you and the University on Dr Woodward's legacy, and am glad you bought the remainder of his collection. I hope this may lay the foundation for enquiries into natural knowledge joined with experiments and observations, and that such studies may be more cultivated daily.[51]

It is clear that collections were secured and information about them disseminated explicitly for the benefit of future generations, and that this link was conceptualised in the eighteenth century. Crucially, though, collections looked backwards as well as forwards, as the antiquarian Thomas Hearne stated to Sloane in 1721,

> I am very sensible of your great Treasure, and, if I should come to London (where I never was yet) I would endeavour to make my self better acquainted with it, especially since there is so much in it about Antiquity. I wish Catalogues of such noble Libraries and Museums as yours were published. Twoud be of great service to Learning, especially if the Owners were, like yourself, of true publick Spirit.[52]

By cultivating studies which helped to connect the "Ancients" with the "Moderns", thus improving the ability to profit from the ever-expanding colonial world, collections were an important part of developing conceptions of

---

50   Hans Sloane, *A Voyage to Jamaica*, Vol. 2 (1725), v; Arnold Hunt, "Under Sloane's Shadow: The Archive of James Petiver," in Vera Keller, Anna Marie Roos and Elizabeth Yale, eds., *Archival Afterlives: Life, Death and Knowledge-Making in Early Modern British Scientific and Medical Archives* (Leiden: Brill, 2018); Richard Coulton, "'What he hath gather'd together shall not be lost': Remembering James Petiver," *Notes and Records of the Royal Society* 74 (2020): 206–207.

51   John Nichols, *Literary Anecdotes of the Eighteenth Century*, Vol. V (London: for the author, 1812), 426.

52   BL, Sloane MS 4046, f.170. Thomas Hearne to Hans Sloane (1 January 1721, Edmund Hall, Oxford).

European society as being on an upwards social, political, economic and intellectual trajectory.

Collecting, preserving and ordering objects (and the knowledge created using them) was understood to not only provide an exemplary record of human accomplishments but also, by their ease of use, provide encouragement to further such achievements and improve knowledge. In this, the collection functioned in the same way as other reference tools in the Enlightenment, such as encyclopaedias or dictionaries, or the catalogues of the collections themselves, all of which were exchanged widely in the name of scholarly progress and, increasingly, national pride.[53] The impulse to identify, define and sort every element of the living world extended into the formulation of powerful and sometimes devastating scientific hierarchies during the eighteenth century, particularly the ways in which non-European societies and races were classified in descending order.[54] In this way, the imagined space became physical: new museums opened across Europe after "first national museum", the British Museum, was established using Sloane's collection in 1753, consolidating not only scholarly taxonomies of the natural and human world, but also the power and authority of the naturalists (and, by extension, nations) who had done the ordering. Compared with the fast-moving, forward-thinking societies of European, armed with science and reason and providence, the rest of the world was represented as being "over there" and "back then."[55]

---

53   Marina Frasca-Spada and Nick Jardine, *Books and the Sciences in History* (Cambridge: University of Cambridge Press, 2000); Richard Yeo, *Encyclopaedic Visions: Scientific Dictionaries and Enlightenment Culture* (Cambridge: University of Cambridge Press, 2001); Rosemary Sweet, *Antiquaries: The Discovery of the Past in Eighteenth-Century Britain* (London: Hambledon and London, 2004); Ann M. Blair, *Too Much to Know: Managing Scholarly Information Before the Modern Age* (New Haven: Yale University Press, 2010); Daniel Margócsy, "'Refer to folio and number': Encyclopedias, the Exchange of Curiosities and Practices of Identification Before Linnaeus," *Journal of the History of Ideas* 71 (2010): 63–89.

54   Francesca Rigotti, "Biology and Society in the Age of Enlightenment," *Journal of the History of Ideas* 47 (1986): 215–233; Londa Schiebinger, "The Anatomy of Difference: Race and Sex in Eighteenth-Century Science," *Eighteenth-Century Studies* 23 (1990): 387–405; Charles W. J. Withers, "Geography, Natural History and the Eighteenth-Century Enlightenment: Putting the World in its Place," *History Workshop Journal* 39 (1995): 136–163; Devin Vartija, "Revisiting Enlightenment Racial Classification: Time and the Question of Human Diversity," *Intellectual History Review* (2020): 1–23.

55   Nicholas Thomas, *Possessions: Indigenous Art/Colonial Culture* (London: Thames and Hudson, 1999); Sadiah Qureshi, "Displaying Sara Baartman, the 'Hottentot Venus'," *History of Science* 42 (2004): 233–257; John M. Mackenzie, *Museums and Empire: Natural History, Human Cultures and Colonial Identities* (Manchester: Manchester University Press, 2009).

In conclusion, the organic interaction of collections and correspondence networks in the early modern period, first on a local and then on a global scale, ultimately helped create an imagined space in which all forms of knowledge were figuratively brought together. Objects were removed from or otherwise stripped of context: through multifarious processes of collection and exchange, through their connection with other forms of knowledge, or quite deliberately and violently as European scholars sought to know, control and profit from a new kind of world. [56] The neutralising space imagined through the interaction of correspondence networks and collections of natural philosophy should therefore be understood as both a by-product of early modern life and a deliberate construction of modern society. The museum collection became an important facet of western scientific objectivity and its cosmopolitan "view from nowhere", providing critical resources for research and representation through the development of the disciplines, as well as for social and political cohesion and control.[57] Thus throughout the history of scientific collecting, and particularly its course during the eighteenth century, we can see how objectivity and authority came to be understood as being owned only by those impartial Western spectators with access to the universal or panoramic view.[58] Indeed, the idea – accepted until only relatively recently, and still difficult for

---

56   Londa Schiebinger and Claudia Swan, eds., *Colonial Botany: Science, Commerce and Politics in the Early Modern World* (Philadelphia: University of Philadelphia Press, 2005); Paula De Vos, "Natural History and the Pursuit of Empire in Eighteenth-Century Spain," *Eighteenth-Century Studies* 40 (2007): 209–239.

57   Tony Bennett, *The Birth of the Museum: History, Theory, Politics* (London: Routledge, 1995); Steven Shapin, "Placing the View from Nowhere: Historical and Sociological Problems in the Location of Science," *Transactions of the Institute of British Geographers* 23 (1998): 5–12; Anne Goldgar, "The British Museum and the Virtual Representation of Culture in the Eighteenth Century," *Albion* 32 (2000): 195–231; Mary Poovey, "The Liberal Civil Subject and the Social in Eighteenth-Century British Moral Philosophy," *Public Culture* 14 (2002): 125–145.

58   Michel Foucault, *The Order of Things: An Archaeology of the Human Sciences* (London: Routledge, 2002); Kim Sloan and Andrew Burnett, *Enlightenment: Discovering the World in the Eighteenth Century* (London: British Museum, 2003); Beth Lord, "Foucault's Museum: Difference, Representation, and Genealogy," *Museums and Society* 4 (2006): 1–14; Alison Byerly, "'A Prodigious Map Beneath His Feet': Virtual Travel and the Panoramic Perspective," *Nineteenth-Century Contexts* 29 (2007), 151–168; Charlotte Bigg, "The Panorama; or, La Nature à Coup d'Oeil," in E. Fiorentini, ed., *Observing Nature-Representing Experience: The Osmotic Dynamics of Romanticism 1800–1850* (Berlin: Reimer, 2007), 73–95; Daniela Bleichmar, "Learning to Look: Visual Expertise Across Art and Science in Eighteenth-Century France," *Eighteenth-Century Studies* 46 (2012): 85–111; Björn Billing, "Circular Visions: Viewing the World from Above in the Late Eighteenth Century," *Journal of Historical Geography* 63 (2019): 61–72.

some to dismiss – that museum collections (along with the practice of science itself) are somehow transcendental and fundamentally objective forms of knowledge is partly a result of the ways in which the imagined space of the collection and its correspondence networks created, controlled and communicated scientific knowledge.

CHAPTER 6

The Dissemination of Chemical Theory and Chemical Instruments through Cabinets, Laboratories, Lecture Theatres and Museums during the Napoleonic Wars

Trevor H. Levere

For Janis Langins, *in memoriam.*

1 Introduction

There are many public and private spaces for science. In museums and universities around the world, but especially in Britain and Europe,[1] there are collections of scientific apparatus and instruments. Sometimes the collections are scattered through numerous departments, but occasionally they are preserved in something like their original groupings, and in rare instances, they are preserved in their original settings. Several of these different spaces and uses may be illustrated by the following instances:

(1) Perhaps most remarkable of all is the museum in Teyler's Foundation in Haarlem, the Netherlands, where the instruments, mostly collected in the Napoleonic era, are still in their original cases, in their original places, and in the original room and building; it is a time-capsule, created by Martinus van Marum.[2] Teyler's also has a library and space for lecturing (including the public performance of experiments) and research, and an art collection; the museum was visited by natural philosophers from around Europe, and by the citizens of Haarlem and neighbouring towns, where they could not only see the instruments, but could also watch lecture demonstrations.

---

1 When I wrote this chapter the European Union still included Britain. Absurdly, by the time this is published, that will no longer be the case.
2 See R.J. Forbes, ed., *Martinus van Marum: Life and Work*, vol. I (Haarlem: Tjeenk Willink, 1969).

(2) Georg Friedrich Parrot[3] built up a cabinet in the new University of Dorpat (now Tartu), used for instruction, but only minimally for research. He asked for Van Marum's assistance in obtaining apparatus.

(3) The Hauch Cabinet in Sorø, Denmark, was assembled by Adam Wilhelm Hauch[4], a distinguished Danish military officer, politician (he became Lord Chancellor), diplomat, and natural philosopher. He used the apparatus and instruments in his cabinet primarily for research. In 1815 he sold his collection to King Frederik, who in 1827 gave it to the Sorø Academy where Hauch's son was a teacher, and became curator of the collection, which was then used in teaching.

As these examples show, the boundaries between cabinets, laboratories, lecture theatres, and museums are not hard and fixed, and nor are the uses made of them; but all of them are spaces of science. In this chapter, I shall consider mainly the apparatus and instruments of chemistry from the years of the chemical and French revolutions to the end of the Napoleonic Wars, and the laboratories and lecture theatres in which they were used. Instruments and apparatus played a key role in the process of suasion and dissemination involved in the chemical revolution. I shall consider some of the key instruments invented by Antoine Laurent Lavoisier, and then show how instruments contributed to Van Marum's conversion to the new chemistry.

Van Marum had bought much of his collection in England, but he went on to design some instruments that were custom-made for him. At first, he hired in-house instrument makers, but the network of instrument makers who supplied his museum and laboratory soon became extensive. A new supplier of instruments for Teyler's was Onderdewijngaart Canzius, the owner of a large instrument manufactory, whose workers were skilled in wood-, glass- and metalwork. After a while, Van Marum came to be frustrated by his in-house instrument makers, and to have sufficient confidence in Canzius to shift the manufacture of key apparatus to the latter's workshop. Van Marum also recommended Canzius to others, among them Georg Parrot, who migrated from France to the Baltic states, then part of the Russian empire, where he was appointed as the first rector and professor of physics (natural philosophy) in the new university of Dorpat (now Tartu).

Parrot's idea of natural philosophy encompassed chemistry, a not unusual pairing. He worked through Van Marum to obtain numerous instruments from Canzius, and his correspondence with Van Marum and with Cuvier, an old

---

3   Georg Parrot (1767–1852). See below.
4   Adam Wilhelm Hauch (1755–1838). Biographical information is at "Adam Wilhelm Hauch," Hauch's Physiske Cabinet, http://www.awhauch.dk/om-hauch/.

friend from school days in Montbéliard, show both his interest in chemistry, and the route by which he sought to obtain its principal instruments. Parrot's experience illustrates the vicissitudes of working in a space almost beyond the periphery of European science in wartime, and the importance of politics for the transmission of chemical and other ideas and matériel. Although today historians of science are more comfortable with networks than with the model of centre-and-periphery, there is no question that Parrot saw Dorpat as peripheral and Paris as central in the world of science, a view shared by Parisians then and perhaps even now.[5]

2   Instruments in the Chemical Revolution

In the introduction to his essay on chemistry in the *Encyclopédie Méthodique* of 1786–1815, Antoine François de Fourcroy drew attention to the technical, instrumental, and manual aspects of pneumatic chemistry.[6] His reason for this was clearly stated. Differences, novelties, and improvements in chemical apparatus had accompanied successive discoveries. New facts might have encouraged the conception and construction of new and improved apparatus, but once that apparatus was used in experiments, it in turn gave rise to new facts. Instruments make possible the controlled production of phenomena. Often designed with particular theories in mind, they can play a reinforcing and self-validating role within the framework of those theories.[7] Fourcroy had thus identified an important aspect of eighteenth-century chemistry, and indeed of

---

5   The model using centre-and-periphery was first used in linguistic theory, dating back to the 1930s, and made its way into historiography by the 1990s. See Elena Fasano Guarini, "Center and Periphery," *The Journal of Modern History* 67 (1995): S74-S96; Antonio García Melmar and José Ramon Bertomeu Sánchez, "Constructing the center from the periphery. Spanish travellers to France at the time of the Chemical Revolution," in Ana Simoes, Ana Carneiro, Maria Paula Diogo, eds., *Travels of Learning. A Geography of Science in Europe* (Dordrecht, Boston, London: Kluwer, 2003), 143–88. That model remains useful, along with network theory. James Delbourgo and Nicholas Dew, eds., *Science and Empire in the Atlantic World* (New York: Routledge, 2008) argue for networks rather than centre and periphery. Both models have heuristic value.

6   Antoine-François Fourcroy, *Encyclopédie méthodique. Chimie, pharmacie et métallurgie* (Paris, 1795–6), vol. III, 262–781, at p. 329. Patrice Bret, "Les chimies de l'Encyclopédie méthodique: une discipline académique en revolution et des traditions de l'atelier," in *L'Encyclopédie méthodique (1782–1832); des Lumières au positive*, eds., Claude Blanckaert and Michel Porret (Geneva: Droz, 2006), 521–51.

7   Ian Hacking, "The Self-Vindication of the Laboratory Sciences," in A. Pickering, ed., *Science as Practice and Culture* (Chicago and London: University of Chicago Press, 1992), 29–64.

experimental science in general. For Lavoisier, as for his contemporaries and immediate predecessors, gas chemistry and the apparatus that made it possible were central. Lavoisier's laboratory had more than six thousand pieces of glassware, many of them used for pneumatic chemistry, and some of them quantitative.[8]

Instruments can serve many purposes, including the persuasion of spectators and participants. It is worth noting that when it came to research Lavoisier often made use of basic, even crude apparatus. As Larry Holmes showed, Lavoisier carried out his seminal research with, at least in the early years, simple and relatively inexpensive apparatus.[9] In his research, he continued to make use of many simple pieces. Subsequent public demonstrations and publications used very different apparatus, which contributed to the rhetorical force of his arguments, in his writings as much as in his demonstrations, and were persuasive arguments for the centring of the Chemical Revolution.

In his textbook and manifesto of that revolution, the *Traité Elémentaire de Chimie* (figure 6.1), Lavoisier stressed the novelty and importance of the large section (roughly one third of the whole), describing instruments and their use:

> In the third part, I have given a description, in detail, of all the operations connected with modern chemistry. I have long thought that a work of this kind was much wanted, and I am convinced that it will not be without use. The method of performing experiments, and particularly those of modern chemistry, is not so generally known as it ought to be; and had I, in the different memoirs which I have presented to the Academy, been more particular in the detail of the manipulations of my experiments, it is probable I should have made myself better understood, and the science might have made a more rapid progress. ... I need hardly mention that third part could not be borrowed from any other work, and that, in the principal articles it contains, I could not derive assistance from any thing but the experiments which I have made myself.[10]

---

8   Levere, "Lavoisier's Gasometer and Others: Research, Control, and Dissemination", in *Lavoisier in Perspective*, ed., Marco Beretta (Munich: Deutsches Museum, 2005), 53–67; Maurice Daumas, "Les appareils d'expérimentation de Lavoisier," *Chymia* 3 (1950): 45–62.

9   Frederic Lawrence Holmes, *Antoine Lavoisier – The Next Crucial Year: Or the Sources of His Quantitative Method in Chemistry* (Princeton, New Jersey: Princeton University Press, 1998), and "The Evolution of Lavoisier's Chemical Apparatus," in Frederic L. Holmes and Trevor H. Levere, eds., *Instruments and Experimentation in the History of Chemistry* (Cambridge, Mass.: MIT Press, 2000), 34–54. Daumas, *op. cit.*

10   Antoine-Laurent Lavoisier, *Elements of Chemistry, in a New Systematic Order, Containing All the Modern Discoveries*, trans. Robert Kerr (Edinburgh: Creech, 1790), xxxiv-xxxv.

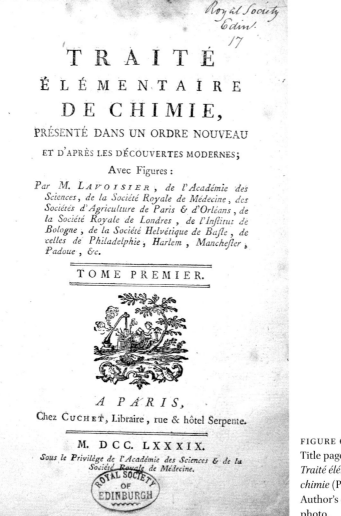

FIGURE 6.1
Title page of Lavoisier,
*Traité élémentaire de
chimie* (Paris, 1789).
Author's collection and
photo.

Lavoisier was not going to give unnecessary credit to his predecessors. In stressing his own originality, he was concerned both with the form and method of
those experiments, and with the novelty of the apparatus used in them, a
novelty that he sometimes exaggerated, but which gave him ownership of the
experiments in which they were used, and of the theories with which they
were associated. The identification of the chemical revolution with the work
of Lavoisier stems partly from his own propaganda and self-promotion, and
partly from the apparatus and instruments that he used in his own laboratory.
Some of the instruments used in key experiments and demonstrations, were

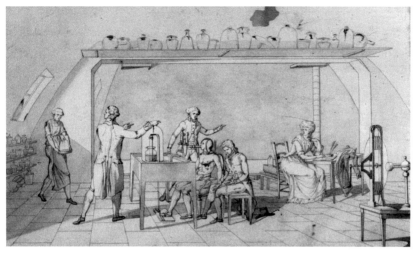

FIGURE 6.2   Experiments on respiration in Lavoisier's laboratory. Mme. Lavoisier drew
the plate and is here shown taking notes. Édouard Grimaux, Lavoisier
1743–1794 (Paris, 1888), facing p. 119. Author's collection and photo.

uniquely available in Lavoisier's laboratory, which reinforced his monopoly on
the chemical revolution. He came to regret that he had not from the outset
stressed the importance of instruments. A careful discussion of instruments
would have helped in ensuring the replicability and acceptability of his results
and would have reinforced the didactic role of laboratory demonstrations.
We can see from sketches made by Mme. Lavoisier that Lavoisier's apparatus
was designed for use in laboratories that were also carefully designed (figure
6.2); we know that chemists visiting Paris were often invited to watch Lavoisier perform key experiments. Seeing was frequently, although not always,
believing.

Most elaborate of all the new pieces of apparatus in Lavoisier's arsenal were
his gasometers (figure 6.3).[11] Weights on a balance pan maintained constant
pressure on gas in the main reservoir. The balance apparatus stood on a col-
umn of solid wood. The balance arm or beam itself pivoted smoothly on a frictionless roller bearing. The rollers in this bearing were made of brass, and the
points supporting the axle or swivel pin of the balance arm were furnished
with bands of rock crystal. The angle of the balance arm could be read using a
brass vernier gauge on a scale graduated in half degrees. From one end of the

---

11   Trevor H. Levere, "Lavoisier's Gasometer and Others. Research, Control, and Dissemina-
        tion," in Marco Beretta, ed., Lavoisier in Perspective (Munich: Deutsches Museum, 2005),
        53–67.

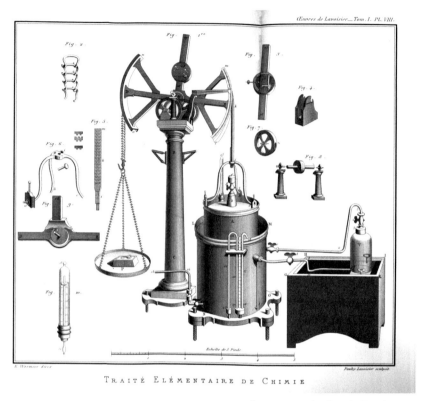

FIGURE 6.3   Lavoisier's gasometer. Lavoisier, *Oeuvres de Lavoisier* vol. 1, (1860): *Traité*
élémentaire de chimie, 2 vols. (Paris, 1789), vol. 1, plate 8. Author's collection
and photo.

balance arm, a scale pan hung, supported by a chain. From the other end, a
Vaucanson chain descended; this chain was built like a modern stainless steel
watch bracelet, flexible without stretching. It was attached to a three-pronged
iron hook, which in turn held an inverted bell jar made of beaten copper,
eighteen inches in diameter and twenty inches high. The bell jar fitted into a
larger copper vessel that was normally filled with water. Pressure of the gas in
the reservoir could be measured by a manometer, and another tube enabled
one to monitor the change in water level during an experiment, and thus the
amount of gas dispensed by the gasometer.

The gasometer was for Lavoisier "a precious instrument, both because of the
large number of ways in which one can apply it, and because there are experiments that are almost impossible to perform without it." He saw it as "what
may be called an *universal instrument*, without which it is hardly possible to
perform most of the very exact experiments." Such special instruments were

expensive – yet another way in which Lavoisier's chemistry was becoming big science. Lavoisier noted:

> In the present advanced state of chemistry, very expensive and complicated instruments are becoming indispensibly necessary for ascertaining the analysis and synthesis of bodies with the requisite precision. ... [I]t is certainly proper to endeavour to simplify these, and to render them less costly; but this ought by no means to be attempted at the expence of their conveniency of application, and much less of their accuracy.[12]

Very few institutions or individuals could afford to equip a laboratory like Lavoisier's.[13] At first, the only way to see demonstrations of the new chemistry was to attend Lavoisier's own performances in his own laboratory. That single laboratory counted as two spaces of science, one for research, the other for public demonstration and suasion. One of the foreign natural philosophers who visited the latter in 1785 was Martinus van Marum, then working in Haarlem.

3 Martinus van Marum: Demonstration, Persuasion, and the Development of Instruments

The Van Marums were originally from Groningen, in the north of the Netherlands. Martinus' father had trained as an engineer and surveyor; he then married and became a master potter in Delft. Martinus was born there, returned to Groningen with his parents, and studied medicine and natural philosophy at the University of Groningen. He qualified as a physician, and in that capacity came to Haarlem in 1776. Eight years later, he became director of the Cabinet of Physical and Natural Curiosities in Teyler's Foundation, which was supported by an endowment provided by a wealthy merchant and philanthropist. A museum was included in the foundation, and Van Marum soon began to build up a cabinet of instruments and apparatus in natural philosophy, stressing

---

12 Daumas, *op. cit* ; Lavoisier, *op. cit.*, pp. 308, 319.
13 There are now two of Lavoisier's gasometers (i.e. instruments constructed broadly on the same plan as the Paris instrument) in the Technisches Museum, Wien. One of these instruments carries a brass plate inscribed "Fortin place de la Sorbonne à Paris 1790" and is of unknown provenance. The other came to the Museum from the k.k. Technische Hochschule in Vienna, but there is no indication of its prior history. I am most grateful to Christian Sichau for information about these instruments.

physical and chemical instruments.[14] His guiding principles in building up the collection were simple — he wanted the biggest and the best,[15] and achieved this strikingly in the case of John Cuthbertson's great electrical machine.[16] He used the instruments for public lectures and demonstrations, as well as for research. Some investigators in the 1780s were exploring the possible connection, perhaps even the identity, of electricity with phlogiston, and so Van Marum soon added chemical experiments to his electrical ones. He travelled to Paris in 1785, where he visited Benjamin Franklin in Passy (now within the bounds of Paris), and told him about his experiments with the powerful new electrical machine. Franklin, American sage-in-residence and honorary Frenchman, had advanced the one-fluid theory of electricity which Van Marum had adopted.[17] He attended lectures and visited eight cabinets of natural history. He also had several discussions with Lavoisier, Gaspard Monge, and Claude Louis Berthollet. Lavoisier invited him to dinner, but Van Marum had too crowded a schedule to accept. Instead, he visited Lavoisier in the Arsenal, and in Lavoisier's laboratory witnessed experiments on the composition of atmospheric air and of water. He was also shown experiments by Berthollet and had exhaustive discussions about chemical airs and the controversies about them. He examined scientific apparatus (especially electrical machines), some in laboratories, some in cabinets. By invitation, he attended meetings of the Académie royale des sciences, at one of which Lavoisier read

14   See G. L'E. Turner and Levere, *Van Marum's Scientific Instruments in Teyler's Museum*: *Martinus van Marum: Life and Work* (Leiden: Noordhoff, 1973), vol. IV.

15   See, e.g., letter of Martinus van Marum to J.H. de Magellan, 9 Nov. 1788, in Roderick W. Home, Isabel M. Malaquias & Manuel F. Thomas, eds., *For the Love of Science: The Correspondence of J.H. de Magellan (1722–1790)* (Bern: Peter Lang, 2017), vol. 2, 1930–1: "Les troubles, qui ont existé quelque temps dans ce pays-ci, ont decouragé les Directeurs de notre Societé Teylerienne, a faire de depenses pour notre museum. Heureusement le repos paroit actuellement bien retabli ce qui les a determiné de suivir le plan (qu'on a fait il y a plusieurs années) de faire un cabinet des instrumens physiques choisies, c.a.d. des meilleurs instrumens, qu'on peut avoir en tout genre, et principalement ceux, qui me peuvent sevir pour des recherches physiques."

16   W.D. Hackmann, "The Design of the Triboelectric Generators of Martinus van Marum, F.R.S.: A Case History of the Interaction between England and Holland in the Field of Instrument Design in the Eighteenth Century," *Notes and Records of the Royal Society of London* 26 (1971): 163–81, and *John and Jonathan Cuthbertson: The Invention and Development of the Eighteenth-Century Plate Electrical Machine* (Leiden: Rijksmuseum voor de Geschiedenis der Natuurwetenschappen, 1973); Lissa Roberts, "Science Becomes Electric: Dutch Interaction with the Electrical Machine during the Eighteenth Century," *Isis* 90 (1999): 640–718.

17   For background and context here, see J.L. Heilbron, *Electricity in the 17th & 18th centuries: A Study of Early Modern Physics* (Berkeley: University of California Press, 1979).

amongst other things, a memorandum on the different sorts of air. But the violent objections which were raised against it, in consequence of which the reading was repeatedly interrupted, and the simultaneous speaking of the Lecturer and his opponents allowed me to hear very little of it.[18]

At another meeting of the Académie, which Van Marum attended, Monge and Jean-Baptiste Le Roy reported on Van Marum's description of his electrical machine, and of the experiments performed with it. "Most of the members listened with extraordinary attention. When the reading was over, I was complimented by many members on the success of my attempts."[19]

By the time he left Paris, Van Marum had seen experiments relating to Lavoisier's new chemical theory and had heard and participated in discussions about them. His confidence in the one-fluid theory of electricity had been strengthened by discussions held and experiments witnessed during his visit. But his confidence in the phlogiston theory had been undermined. The arguments and practical demonstrations of the French chemists did not immediately convince him, so he set about a careful and protracted examination of the experimental evidence for the new antiphlogistic theory. He repeated some of his chemical experiments using Cuthbertson's electrical machine in Teyler's Foundation, the crown jewel of the cabinet of natural philosophy and chemistry (figure 6.4). Experiments were essential; Van Marum, like the early members of the Royal Society of London, could happily have adopted that Society's motto, *Nullius in Verba*. Where Van Marum lacked the necessary apparatus, he had it built. By the of the end of the winter of 1785–86, he had come around to the new French theory.[20] He soon afterwards published an outline of Lavoisier's chemistry,[21] although he was still doubtful about parts of it, notably those concerning caloric. This was the first comprehensive statement of the new theory, and antedated Lavoisier's own announcement in his *Traité* by two years. Van Marum was criticized for his ready phlogistic apostasy by Dutch chemists who had not made the scientific pilgrimage to Paris. He set about converting them to the new theory. Talk was not enough. Practical demonstrations were

---

18 Van Marum, "Journal Physique de Mon Sejour à Paris 1785", in R.J. Forbes, ed., *Martinus van Marum Life and Work*, vol. II (Haarlem: H.D. Tjeenk Willink, 1970), 220–29 at p. 222.

19 Ibid., p. 236.

20 Levere, "Martinus van Marum and the Introduction of Lavoisier's Chemistry in the Netherlands," in *Martinus van Marum*, vol. I, 158–286.

21 Martinus van Marum, *Schets der Leere van M. Lavoisier 1787. Facsimile* (Delft: Koninklijke Nederlandse Chemische Vereniging, 1987). Levere, op. cit.; See also Levere, "The Interaction of Ideas and Instruments in Van Marum's Work on Chemistry and Electricity," *Martinus van Marum*, vol. IV, 103–23.

FIGURE 6.4   The oval room, Teylers Museum, Haarlem. Oil painting on wood by Wybrand
Hendriks, ca. 1810. The only significant change to have occurred in the
museum and cabinet since the early 19th century is that the great electrical machine, shown here front and centre, encased, has been moved to the
19th-century room. Photo taken by Teylers Museum. Courtesy of Wikimedia
Commons, Public Domain, License CC BY-SA-3.0.

needed, and required apparatus was not yet available outside Paris. This was
quite clear to Van Marum. He remarked that

> one of the principal reasons why [Lavoisier's] theory formerly enjoyed
> such slight attention among Physicists and Chemists in this country
> seemed to me the fact that they had no opportunity to see or to repeat the
> experiments, the results of which formed the basic principles of the new
> chemical theory. Indeed, the necessary apparatus as made by the generous
> Lavoisier at his own expense could hardly be obtained, owing to its costli-
> ness and to the difficulty of constructing it with the precision required. This
> consideration suggested to me that I could contribute to the progress of the
> Lavoisierian chemistry ... by making in Holland some of the most impor-
> tant experiments, on the results of which this theory is largely founded.[22]

22   Van Marum, trans. J.H. Zorn, *Description of Some New or Perfected Chemical Instruments
Belonging to Teyler's Foundation, and of Experiments Carried Out With These Instruments*
(Haarlem, 1798), in *Martinus van Marum*, vol. V, 241.

The demonstration of the composition of water by the continuous combustion of hydrogen gas with oxygen gas struck him, as it did others, as the most important in this process of persuasion. But Lavoisier's gasometers, required for this demonstration, were prohibitively expensive, and Van Marum wanted to see if he couldn't do the job with cheaper but no less effective forms of gasometer.

The funds made available through Teyler's Foundation enabled him to pursue his enquiries. At first, he met with frustration. Five years were to pass before he had the necessary apparatus and was able to achieve consistent experimental results. In 1791, he publicly performed the composition experiment in Haarlem for several days on end. His demonstrations, carried out in Teyler's Foundation with instruments displayed in Teyler's Museum, won converts. The museum and adjacent laboratory served as spaces for public demonstration and private research. Van Marum was also delighted to find that his relatively inexpensive gasometers were ordered from Dutch instrument makers and copied widely. The construction was simple. Instead of the sophisticated frictionless bearings, vernier gauges, and the system of counterweights used by Lavoisier, Van Marum used the constant pressure of a head of water to control the rate of gas flow (figure 6.5). He sent two letters to Berthollet explaining his

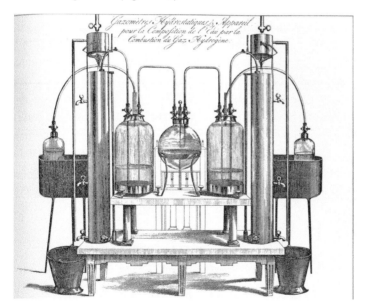

FIGURE 6.5   Van Marum's hydrostatic gasometer: Apparatus for the combustion of hydrogen and oxygen to form water. Van Marum, *Verhandelingen, uitgegeeven door Teyler's Tweede Genootschap,* 10: *Beschryving van eenige nieuwe of verbeterde chemische werktuigen behoorende aan Teyler's Stichting, en van de proefneemingen met dezelve in 't werk gesteld* (Haarlem: Johannes Jacobus Beets, 1798) IV, Plate I. Courtesy of Teyler's Museum.

design, and the letters were published first in the *Annales de Chimie* of 1792, and then in several German and other journals.[23] The demonstration of the composition of water effectively required four hydrostatic gasometers, two as reservoirs supplying the other two constant pressure gasometers; but this was still a lot cheaper than Lavoisier's apparatus. Van Marum also developed an even simpler arrangement using just two gasometers, as well as apparatus for the combustion of phosphorus, of charcoal, of oil, and of most of the principal experiments or demonstrations described in Lavoisier's *Traité*. Van Marum's instruments were described in 1798, in his *Description of Some New or Improved Chemical Instruments*. He kept his account tolerably concise by insisting that readers should first become familiar with Lavoisier's books and papers.[24]

There were problems to overcome in building the new apparatus, some arising from dealing with instrument makers,[25] and others from the political and military circumstances of the day. Even before the French Revolution, Dutch society had been divided between supporters of the House of Orange and would-be republican burgher-regents. The latter appropriated to themselves the title of Patriots. In 1787, an insult to prince William of Orange, who was married to the sister of the Prussian King, provoked a Prussian invasion. The Patriots were soon removed from positions of political power. Following the French Revolution, the political polarization of the Netherlands again came to a head, supporters of the House of Orange deploring plebeian power, while the Patriots were sympathetic towards the Jacobins. Still, the Netherlands tried to remain neutral. When Dumouriez took the Austrian Netherlands in 1792, the Dutch United Provinces could no longer keep out of the war. Political uncertainty translated into economic troubles, and the Directors of Teyler's Foundation were forced to rein in their expenditures, thus slowing down Van Marum's research program. Getting scientific supplies and material such as glassware and platinum wire from abroad also became difficult.[26] By 1795, anti-Orangist sentiment and philo-Jacobinism had together grown to the point where the

---

23   Van Marum, "Lettre à Berthollet, contenant la description d'un Gazomètre, ... et d'un appareil pour faire ... l'expérience de la combustion de l'eau, par combustion continuelle, avec plus de facilité et moins de frais," *Annales de Chimie* 12 (1792) : 113–40; "Seconde lettre à Berthollet ...," *Annales de Chimie* 14 (1792) : 313–23. Full references for these and the German publication are in J. G. de Bruin, "Van Marum Bibliography," in *Martinus van Marum*, vol. I, 287–320 at 293–4.

24   For the antiphlogistic campaign, the letters to Berthollet and a brief account of Van Marum's simplified gasometers, see Levere (1969), 195–201, 222–9.

25   See below.

26   See, e.g., *Minutes of the Directors of Teyler's Stichting*, Teyler's Foundation MSS, vol. I, p. 276, minutes of 10 Oct. 1794.

French army could take advantage of the situation to liberate their Dutch Patriot brethren. French forces crossed the frozen rivers and entered Holland from the south. William V, unwilling to face the grim realities of liberty, equality, and fraternity, fled with his family to England, and his nation was transformed into the Batavian republic. Dutch philo-Jacobins were elevated to senior administrative posts and were helped by French authorities to reorganize the country. In 1806 the Batavian Republic morphed into the Kingdom of Holland, ruled by the newly crowned and gentle King Louis Napoléon Bonaparte (born Luigi Bounaparte), a younger brother of Napoleon I, Emperor of France. Van Marum's friendship with French scientists, most of whom, unlike Lavoisier, survived the Terror, as well as a very skillful interview in 1806 with King Louis, enabled him to preserve Teyler's Museum, and to give public lectures and to pursue his now supposedly useful research, the only kind countenanced in the new political order.[27]

Teyler's had a succession of instrument makers in these turbulent years. Friedrich Wilhelm Fries, born in Strasbourg and working in London, was appointed to Teyler's in 1790.[28] He did valuable work on gasometers and other chemical apparatus, but fell out with Van Marum after a couple of years. A good deal of work was then done for Van Marum by Amsterdam craftsmen, and then gradually Van Marum gave more and more work to Onderdewijngaart Canzius,[29] a former lawyer whose allegiance to the House of Orange cost him his practice when the French came. Canzius thereupon changed his profession, and in 1797 set up a factory for manufacturing scientific instruments, at a time when competition from the English was erased by the military situation. By around 1800, Van Marum was relying on him as one of the principal sources, perhaps the principal source of instruments for Teyler's.

4   Onderdewijngaart Canzius and the Peter Principle

Canzius began in Brussels, then set up a factory in Delft. He employed local craftsmen, and also brought in instrument makers from Germany. By 1798 he had a staff of 30, and two years later his complex operations involved

27   Levere (1973), 76–7, 85–6.
28   Ibid., 65–7.
29   Biographical information about Canzius is from Peter de Clercq, "J.H. Onderdewijngaart Canzius, Instrument Manufacturer and Museum Director," *Bulletin of the Scientific Instrument Society* 49 (1996): 22–4. See also de Clercq, "The Instruments of Science: The Market and the Makers", in Klass van Berkel, Albert van Helden and Lodewijk Palm, eds., *A History of Science in the Netherlands: Survey, Themes and Reference* (Leiden: Brill, 1999), 311–31.

seventeen departments for copperwork, carpentry, glassware, etc. In 1804 his catalogue listed some 650 items, including mathematical, physical, chemical, constructional and mechanical instruments, as well as surgical tools, musical instruments, anatomical preparations, and a whole host more – a quite extraordinary range.[30]

His factory flourished, and he was rewarded not only with profits, but also with membership in philosophical societies. When the Republic became a Kingdom, Canzius thrived in the new order. His entries at the 1808 Utrecht exhibition of national industry received a gold medal, and he was soon supplying King Louis with instruments. Van Marum judged that the dividing engine from Canzius's "factory was no good, and in consequence almost everything is bad for which it has been used. However, the physical and chemical instruments, – insofar I have seen them – were very good." Even so, "that factory, through no fault of his, but as a result of the great enterprises King Louis demanded from him, was thrown into disorder and came to an end."[31] Canzius became over-extended in meeting the ambitious orders that Louis sent him. His factory failed in 1810, the year in which Louis had to abdicate for being too loyal to his Dutch subjects. Canzius was forced to dissolve and sell his enterprise. In the same year, Holland was absorbed into the French empire, until it successfully revolted after the defeat of Leipzig in 1813. The Prince of Orange, son of the exiled William V, was recalled. In 1815 he was crowned William I, King of the Dutch and Belgian provinces, until the Belgian revolt fifteen years later.

Van Marum later (in 1827) told his friend Gerard Moll in Utrecht that Canzius

> specialized ... in the knowledge of many factories and everything concerning industry, as became evident to me on the occasion of the exhibition at Ghent, and afterwards here as well. On account of the knowledge and activity, shown at Ghent, the Minister appointed him before the last exhibition to direct everything that would be needed for its organization, and His Majesty [William I], being informed about his merits before and during the last exhibition, has since granted him the title of Referendaris. All this was the reason why he was appointed by His Majesty last winter as Director of the National Museum of Art and Industry at Brussels.[32]

---

30   *Catalogus van Mathematische, Physische, Anatomische, Chirurgische en andere Instrumenten, te bekomen in de Fabricq van Mr. J. H. Onderdewijngaart Canzius ..., te Delft* (1804).

31   Van Marum to Moll, 28 May 1827, translation in *Martinus van Marum. Life and Work*, vol. VI, 242.

32   Ibid., 242–3. Moll had a less enthusiastic view than Van Marum about the quality of Canzius' instruments: Moll to Van Marum, 29 May 1827, ibid. 243. "You know my opinion

Nothing succeeds like failure. But while Canzius's factory was thriving, in the years of the Batavian republic and at the beginning of Louis' reign, he made and sold most of the chemical instruments devised by Van Marum. It was natural then, when Van Marum received a request from Parrot in Dorpat to supply him with chemical and electrical instruments, that Van Marum would refer him to Canzius.

5   Georges Parrot and the Establishment of a Philosophical Cabinet and a Chemical Laboratory in Dorpat

5.1   *Chemistry in a New University*

Some years ago one of my colleagues, Janis Langins, came back from a conference in the Baltic states with a small catalogue of physical instruments in the Museum of the History of Tartu State University in Estonia.[33] The instruments were part of a *cabinet de physique* from the first two decades of the nineteenth century, and they came from a wide range of European and local sources. The suppliers included George Adams in London, whose workshop had provided some of the choicest pieces for King George III's collection;[34] J.H. Tiedemann in Stuttgart;[35] local instrument makers to the university, including Baron Christoph von Welling and his successor the watchmaker Politour;[36] and the Netherlands manufacturer J.H. Onderdewijngaart Canzius,[37] who had had factories in Brussels and Delft, and who had been a principal supplier for Martinus van Marum and Teyler's Foundation in Haarlem.[38] The collection in Tartu once comprised around 450 instruments, of which only a minority have

---

about the instruments of O. W. C. If only I do not have to use them, I have nothing against the man himself."

33   Erna Kõiv, *Tartu Ülikooli ajaloo muuseumis. Kataloog* (Tartu: Tartu Ülikool, 1989).

34   John R. Millburn, *Adams of Fleet Street, Instrument Makers to King George III* (Aldershot and Burlington: Ashgate, 2000). Alan Q. Morton and Jane A. Wess, *Public & Private Science: The King George III Collection* (Oxford: Oxford University Press in association with the Science Museum, 1993).

35   Johann Heinrich Tiedemann (1752–1811). See Andor Trierenberg, *Die Hof- und Universitätsmechaniker in Württemberg im frühen 19. Jahrhundert* (Dr. Phil. dissertation, Stuttgart University, 2013), 272–83.

36   Benjamin Politour, b. ca. 1774, instrument maker to the University, watchmaker. I have found no information about von Welling as instrument maker.

37   Peter de Clercq, "J.H. Onderdewijngaart Canzius, Instrument Manufacturer and Museum Director," *Bulletin of the Scientific Instrument Society* 49 (1996): 22–4.

38   *Martinus van Marum. Life and Work*, 6 vols. (vols. 1–3, Haarlem, 1969–73, ed. R.J. Forbes; vols. 4–6, Leiden, 1973–76, ed J.G. de Bruijn).

survived the ravages of time and war. The creator of the cabinet was Georges Parrot,[39] born in Montbéliard near the borders of France, Switzerland, and Germany. A good part of the original collection was designed to demonstrate or illustrate key aspects of Lavoisier's new chemistry which, by 1802, when Parrot's university was founded,[40] had acquired general but neither universal nor unqualified assent.

Montbéliard, Parrot's hometown, had become part of the house of Württemberg in 1397. It took four hundred years of French invasions and harassment before the town was made part of the French Republic in 1793. Parrot, like many Montbéliards, had to leave home to seek a career. He was the youngest of eighteen children of a physician and one-time mayor of Montbéliard. He studied at the Karlschule in Stuttgart, a training school for bureaucrats controlled by the Duke and offering a good education in the sciences. From Stuttgart, he went as tutor to a noble Protestant Norman family, came back to Germany in 1788, married, and was soon widowed. In 1795 he left for the Baltic provinces of the Russian empire as tutor in the household of a liberal Baltic nobleman, Count Karl von Sivers. The nobles in Livonia were interested in improving agriculture and their estates, and to that end founded the Livonian Public Utility and Economic Society, a Baltic analogue of the Royal Institution in London. Parrot was the first secretary. He pursued an extremely wide range of scientific studies, in chemistry, meteorology, physiology, terrestrial magnetism, upper atmospheric phenomena such as aurora borealis, instrumentation, geology, natural history, and physics. Parrot at first felt thoroughly isolated in the Baltics, which he described in a letter to his compatriot Cuvier as a "boreal cavern." Nevertheless, he had been studying modern chemistry, publishing papers from time to time. He had recently published a paper in Scherer's journal, offering a new theory of the nature of carbon that would explain "several phenomena of vegetation in a very simple fashion. ... Do see, my dear friend, if these labours can't get me employment in Paris." He believed that this work would force a new interpretation of Lavoisier's theory in relation to plant and animal chemistry."[41]

---

39   Biographical information on Parrot is from Janis Langins, "Diverging Parallel Lives in Science: Unpublished Correspondence from Georges-Frédéric Parrot to Georges Cuvier," *Journal of Baltic Studies* 35 (2004): 297–320, and in Roderick von Engelhardt, *Die Deutsche Universität Dorpat in ihrer geistesgeschichtlichen Bedeutung* (Reval, 1993), 24–47 and 176–179.

40   Hugo Semel, ed., *Die Universität Dorpat (1802–1918). Skizzen zu ihrer Geschichte* (Dorpat: H. Laakmann, 1918), 8–13, gives an account of Dorpat University and Parrot's role therein.

41   Parrot to Cuvier, n.p., n.d. [Riga, 1801], Cuvier MS 223, no. 46, Bibliothèque de l'Institut, Paris, published in Langins, op. cit., 311–2. Parrot and apothecary Grindel of Riga, "Ueber

Parrot's isolation was political and cultural as well as geographical. In 1796, the year after Parrot's arrival in the Baltics, the reactionary Tsar Paul I succeeded his mother Catherine the Great and set about cutting off Russia from destabilizing European and especially French ideas. He planned to establish a new university for the Baltic provinces in Mitau[42] to keep his subjects at home, and, by giving the German Baltic nobility a German Protestant University, to compensate them for the lack of foreign travel – they needed express permission to leave Russia. Tsar Paul was assassinated on 24 March 1801. Alexander succeeded him as Tsar and changed the location of the new university to Dorpat[43] where the university opened in 1802.[44] Dorpat had a chequered history. Founded by the Teutonic Knights in 1403, it had been a Hanseatic city in the Renaissance, was then taken over by Poland, and next became part of the Russian empire. Dorpat was near St. Petersburg and Tsar Alexander at least began his reign as a liberal. Parrot was able to get the Imperial ear – indeed, he was at Alexander's beck and call, and had frequent private audiences with him. He was put in charge of the schools in the four provinces of his department. He was also the first Rector and professor of physics at the University of Dorpat. He held the post of Rector from 1802 to 1803, and again from 1805 to 1806, and from 1812 to 1813. He held the chair of physics from 1802 until 1826, when he became a full member of the Petersburg Academy of Science. Imbued with the ideals of the Enlightenment, he was able to prevent the university from coming under the control of conservative members of the nobility. Thanks to Parrot's efforts, Dorpat achieved a reputation that was to spread beyond the borders of the Russian empire. Dorpat's students and faculty have included the astronomer and geodesist Friedrich Georg Wilhelm von Struve, the physicist Heinrich Friedrich Emil Lenz, the naturalist and explorer Karl Ernst von Baer, and Friedrich Wilhelm Ostwald, one of the founders of physical chemistry and an early Nobel laureate.

It was not as Rector but as professor of physics that Parrot wrote about his electrical and galvanic research to Van Marum, and then in 1803 asked him for "a service which you have offered to all natural philosophers, that of sending me, or rather having made under your supervision your large and your small gasometer. These are for our university, for which I am putting together

die reine Kohle," *Allgeimenes Journal der Chemie* 7 (1801): 3–8. Langins tentatively dates Parrot's letter to 1800, but the publication in Scherer's *Journal* suggests a date of 1801.

42   Now Jelgava near Riga, in Latvia.
43   Now Tartu, in Estonia.
44   Semel, ed., *Die Universität Dorpat .... Skizzen zu ihrer Geschichte* 8–13, gives an account of Dorpat University and Parrot's role therein.

a complete set of physical apparatus." If the war continued until the opening of northern navigation in spring, he asked Van Marum to have the instruments sent to Lübeck, whence they could be shipped to Riga or Revel (the modern Tallin) or another Baltic port.

Van Marum replied by saying that Canzius now made all such instruments for him:

> Formerly I have had made here under my supervision, by different crafts-men, the physical and chemical instruments of which I have given a description. But since Mr. Onderdewijngaart Canzius started an excellent factory of physical and chemical instruments in Delft, 7 hours' journey from here, where one can get made in the very best way whatever one desires, I let him make everything I need, be it for my own use of for Tey-ler's Museum; and also all that the foreign scientists ask me to get copied from what I have described.

Van Marum praised Canzius's work, told Parrot that he often had instruments ready for dispatch, and urged Parrot to write directly to him. Van Marum had already asked Canzius to send Parrot a catalogue. He undertook to inspect any instruments that Parrot ordered. In return, Van Marum asked Parrot to find him a source of Siberian minerals, and a large and powerful lodestone.[45]

Parrot ordered a variety of apparatus, including Van Marum's large gasom-eter, his small gasometer, the apparatus for the combustion of oils, and other apparatus for the combustion of alcohol, the oxidation of mercury, the for-mation of phosphoric acid, and the formation of carbonic acid. A year later, nothing had yet been sent, and Parrot had not had any explanatory letters from Canzius. In May 1804, Van Marum promised that as soon as the instruments were ready, he would go to Delft to inspect them. A year further on, with no progress, Parrot wrote to Van Marum about the room for the Cabinet de Phy-sique, with an adjoining laboratory, and nearby an astronomical observatory. In 1806, Canzius had made most of the apparatus, but reported that he had encountered problems in shipping goods from either Rotterdam or Amster-dam. The Napoleonic Wars weren't helping. All the apparatus was at last ready and packed and awaiting the first ship leaving for Russia, thus going all the way by sea rather than first going overland to Lübeck.

When the instruments finally arrived, in 1809, Parrot promptly arranged for payment to be made through a banker in St. Petersburg. Wartime problems of

---

45   Van Marum to Parrot, 29 Nov. 1803, in *Van Marum. Life and Work*, vol. VI, 271–2 at 272. Par-rot obliged. The handsomely encased lodestone is shown in Turner and Levere, *Martinus van Marum*, vol. IV, 188–9.

communication meant that even the transmission of money was a problem, and it took a while to get the funds to Canzius. Meanwhile, Parrot reported, the large gasometer was exactly as Van Marum had described it, but he couldn't use it because one of the large jars arrived in pieces. The glass workers in Petersburg couldn't make anything of that size, and so Parrot, after six years of delay, had to write to Canzius again to order another jar and "several other items, as much to replace what has been broken [in transit] as to get new things." A year later, Parrot was asking Van Marum to get Canzius to send him the replacement glassware: "I consider that it is his obligation, since I paid the insurance." But 1810 was the year in which Canzius's factory went bankrupt. Van Marum wrote sadly that, in Holland's wretched state, the factory couldn't recover, and so he could no longer help Parrot.[46]

Under the circumstances, it is remarkable that Parrot received any instruments at all; it is also remarkable that a fair number of them are still in the university that he helped to create. Thanks to him, the instruments for the propagation of the chemical revolution reached the Baltics in 1810.

5.2   *Adam Wilhelm Hauch, Chemical Apparatus and the Chemical Revolution in Denmark*

The Danish soldier, statesman and natural philosopher Adam Wilhelm Hauch obtained his introduction to chemistry from his friend P.C. Abildgaard, Professor at the Veterinary School, and from Professor C.G. Kratzenstein of the University of Copenhagen. In 1786 he went on leave from the Court in Copenhagen to concentrate on natural philosophy. In 1788–1789, the year leading to the publication of Lavoisier's *Traité*, and also to the French Revolution, Hauch went on a scientific tour, visiting Klaproth in Berlin, Van Marum in Haarlem, Priestley and Cavendish in Birmingham and London, and Lavoisier in Paris.[47] When he returned, Hauch began to build his collection of apparatus, and from 1790 he gave demonstration lectures using instruments from the collection. Instruments carried educational, rhetorical, and suasive power. Whether for public lectures or for private performances before a select audience, instruments were essential. Hauch also used them for research, including work on the composition of water that he presented to the Royal Danish Academy of Science in 1791, and, in modified form, to the Royal Society of London in 1793. His researches were published not only in Denmark, but also in German journals. In 1794 he

---

46   Correspondence between Van Marum and Parrot from 1804–1810, in *ibid.*, pp. 272–5.
47   Martin Heinrich Klaproth (1743–1817), the first eminent German chemist to adopt Lavoisier's chemistry. Joseph Priestley (1733–1804), dissenter in religion and politics, who adhered to the phlogiston theory until his death. Henry Cavendish (1731–1810), chemist and natural philosopher.

published a textbook based on his lectures,[48] presenting chemistry in terms of Lavoisier's antiphlogistic system. He remained active in science until 1801, when he returned fully to court duties. He published a beautifully illustrated first part of his description of his physical cabinet in retirement, when he had turned eighty, just two years before his death in 1838.

The Hauch cabinet, like the Van Marum collection in Teyler's and cabinet assembled by Georges Parrot in Dorpat, was assembled as a comprehensive cabinet in natural philosophy.[49] For those who, like Hauch, adopted Lavoisier's theory and promoted the chemical revolution, any effective cabinet had to have the instruments essential for repeating and demonstrating the key experiments in Lavoisier's *Traité*. We have seen the centrality of gasometers. Lavoisier's made heavy demands on the instrument maker's skill and were prohibitively expensive. This was a major problem in Denmark, which was not a rich country in the 1790s; with England's help, Denmark, in 1807, was headed for bankruptcy as a victim of the Napoleonic Wars.[50]

Part of Hauch's challenge, then, like Van Marum's, was to find cheaper solutions to problems posed by Lavoisier. He invented minor improvements to some of Lavoisier's apparatus, as was the case with his filter funnel (figure 6.6); he made major modifications, for example using box-shaped gasometers (figure 6.7), which were easier and cheaper to build even than Van Marum's, although he subsequently bought Van Marum's gasometers manufactured by Canzius; he devised new instruments, for example his eudiometer; and in those cases where Lavoisier's apparatus was both affordable and efficient, for example his ice calorimeter (figure 6.8), he either bought copies or had them made locally.

---

48   Hauch, *Begyndelses-Grunde til Naturlæren*, 2 vols. ([Copenhagen],1794), German trans-
      lation *Anfangsgründe der Experimental-Physik* (Schleswig, 1795–6). Hauch's publications
      are listed in "Bibliografi," Hauchs Physiske Cabinet, http://www.awhauch.dk/litteratur/

49   Anja Skaar Jacobsen, "A.W. Hauch's Role in the Introduction of Antiphlogistic Chemistry
      into Denmark," Ambix 47 (2000): 71–95; Levere, "The Hauch Cabinet: Chemical Appara-
      tus and the Chemical Revolution," Bulletin of the Scientific Instrument Society 60 (1999):
      11–15; Ole Bostrup, *Dansk kemi 1770–1807: den kemiske revolution* (Copenhagen, 1996).
      There is a well-illustrated online catalogue of the Hauch collection, organized by sub-
      ject: "Instrumenter fra Hauchs Physiske Cabinet," Hauchs Physiske Cabinet, http://www
      .awhauch.dk/instrumentkatalog/kategori/

50   Denmark was at first neutral during the French wars but was under strong pressure from
      France and Russia to align itself with Russia. Denmark tried to remain neutral, but to
      pre-empt any move of Denmark to join the French alliance, the British Royal Navy sank
      much of the Danish fleet at harbour and captured the remaining ships. For good measure,
      they heavily bombarded Copenhagen. One result of this action was that Denmark for-
      mally allied itself with France, but since it now had no fleet, its value to that alliance
      was small; and since Denmark depended on its ships for trade and fishing, its economy
      rapidly tanked.

FIGURE 6.6
Filter funnel, hauch cabinet.
Photo by the author.

There are also whole categories of instruments included in the Hauch cabinet
that owe nothing to the chemical revolution. They are instead the product of
investigations with a quite separate history, that only coincidentally converged
with the work of Lavoisier and his colleagues. A fine example of this category
is the blowpipe,[51] that was used in antiquity in the manufacture of jewelry,
continued in use through the work of alchemists and metallurgists, and in the
late eighteenth century underwent a renaissance in the hands of mineralogists,
whence it re-entered the mainstream of chemistry. Finally, we need to remem-
ber that although Hauch's activity in chemical research was concentrated in
the years of the promulgation of the chemical revolution, he made additions to

---

51   See e.g. J. J. Berzelius, *The Use of the Blowpipe in Chemical Analysis, and in the Examination
      of Minerals*, trans. J. G. Children (London: Baldwin, Cradock, and Joy, 1822).

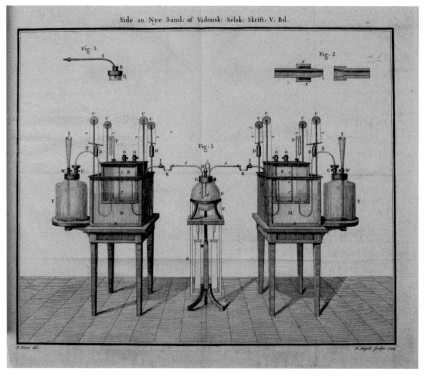

FIGURE 6.7   Hauch's gasometers and combustion globe for experiments on the composition
of water. A.W. Hauch, "Beskrivelse af en forbedret Gazometer eller Luftmaaler,
og nogle med samme anstillede Forsøg," Nye Samling af det KongeligeDanske
Videnskabernes Selskabs Skrifter, 5, 1799, 18–27, facing p. 20. Biodiversity
Heritage Library 180233.

the cabinet over ensuing decades, so that Liebig's *Kaliapparat* of 1831,[52] which
revolutionized the practice of organic analysis, appears in Hauch's 1836 description of his cabinet. Many of the chemical instruments in the cabinet were for
the demonstration of Lavoisier's key experiments, but there is much besides.

Heat is the principal agent of chemical change, and so apparatus for the
controlled application of heat is of crucial importance. There are two major
categories: muffle or chamber furnaces, where the heat is applied indirectly,
the substance under investigation being placed in a chamber which is heated
by the products of combustion; and furnaces or lamps where heat is applied
directly. The use of retorts and crucibles in combination with instruments of
the latter category provides another way to keep the combustion products
separate from the material being heated. Hauch had a good range of chamber

52   Melvyn Usselman, Alan Rocke, Christina Reihart and Kelly Foulser, "Restaging Liebig: A
Study in the Replication of Experiments," *Annals of Science* 62 (2005): 1–55.

FIGURE 6.8
Lavoisier and Laplace's
ice calorimeter. Hauch
cabinet. Photo by the
author.

ovens, and of sources for the direct application of heat, including a muffle oven, a spirit lamp devised by Berzelius, a more elaborate chemical oven, and a spirit lamp attributed by Hauch to Guyton de Morveau, which uses an Argand lamp.[53]

---

53   The instruments referred to in the following pages are illustrated in the online catalogue, "Instrumenter fra Hauchs Physiske Cabinet," Hauchs Physiske Cabinet, http://www. awhauch.dk/instrumentkatalog/kategori/ . A general account is given in H. Andersen, "Fysik og Museum: Hauchs Physiske Cabinet," *Nordisk Museologi* 2 (1994): 31–46. Hauch's

Apparatus for distillation could be simple or complex, and junctions between different parts of the apparatus were sometimes effected by lutes. Lutes were likely to contaminate the reactants, often leaked, and sometimes could not be removed without destroying the apparatus. Lavoisier and his co-workers and instrument makers had made extensive use of brass couplings, lacquered to the glassware, and bearing screw threads. Rubber tubing was not available during the chemical revolution; extruded glass tubing was becoming available, and Hauch made extensive use of it. Ground glass connections were expensive, since in the 1790s and 1800s each joint had to be ground individually, a process at once time consuming and demanding skill. It is not surprising that few of Hauch's pieces involve ground glass connections.

He modified a variety of standard apparatus. Vessels for the combustion of hydrogen and oxygen had not changed radically from those used by Lavoisier and by Van Marum. The combustion chamber in the Hauch cabinet is like the one devised by Gaspard Monge, a glass sphere with a brass cover, through which tubes and electrical wires were set (figure 6.9). The gases enter the sphere, and are ignited by an electric spark, produced invariably, at least before 1800, by a frictional electrical machine, for example a cylindrical machine, like the one in the Hauch cabinet. But gasometers were another matter – Lavoisier's were too expensive to buy, and extremely demanding to make. There were two routes open to Hauch, to import simpler and cheaper ones, or to have simpler and cheaper ones made locally to his own specifications. He took both of these routes. He bought at least one of Van Marum's gasometers, made by Canzius. It is probable that local glassworkers were unable to produce jars of sufficient size, so that ordering one from abroad may have been Hauch's only option.

Hauch also devised his own gasometer, which might more properly be termed a modification of the gasometer made by Dumotiez in Paris. This appears to have been based on one of Lavoisier's early ideas for a gasometer; it maintains constant pressure on the gas to be stored and transmitted, by means of a system of weights and pulleys. Dumotiez's instrument used a bell jar, similar in scale and appearance to Van Marum's. Presumably because of the difficulties of making such a jar, Hauch substituted a box with sheet glass sides for the jar but kept the pulley system used by Dumotiez. Hauch used his own gasometers for his work on the composition of water in 1793. It is worth noting

------

own account is A.W. Hauch, *Det Physiske Cabinet eller Beskrivelse over de til Eksperimental-Physiken henhorendevigtigste Instrumenter tilligemed Brugen deraf,* 2 vols. (Copenhagen, 1836).

FIGURE 6.9
Combustion globe, after
Monge. Hauch cabinet.
Photo by the author.

that Parrot's son Friedrich, who wrote a survey of gasometers,[54] was favourably impressed by Hauch's design. But the aftermath of the French Revolution was not the best time for disseminating inventions from a small uneasily neutral nation. By the time the Napoleonic Wars were over, gasometers had been displaced by cheap and simple gas holders.

After the 1790s, Hauch appears to have carried out no significant chemical work, although he added to his cabinet. Chemistry may have been the loser, but his cabinet, like Van Marum's in Haarlem and Parrot's in Dorpat, provides us with a time capsule in the spaces occupied by the instruments of chemistry, a unique record of that science in the turbulent years of the French Revolution, the Napoleonic Wars, and the chemical revolution.

---

54   Friedrich Parrot, *Ueber Gasometrie nebst einigen Versuchen über die Verschiebbarkeit der Gase* (Dorpat: M.G. Grenzius, 1811).

Acknowledgements

Long ago, R.J. Forbes introduced me to van Marum and the instruments at Teyler's Museum, where J.G. de Bruijn was especially helpful. G.L'E. Turner then joined me in a study of Teyler's Museum and its scientific and philosophical instruments. Janis Langins gave me transcripts of correspondence between Parrot and Cuvier, which he subsequently published, and he was a valued guide in introducing me to the history of the University of Tartu (then Dorpat). Erna Kõiv, who compiled the catalogue of instruments in the University Museum, was generous in answering queries. My greatest debt incurred while working on the Hauch Cabinet is to Jørgen From Andersen and his wife Helle, who welcomed me into their home for a week. Jørgen was the curator of that cabinet and helped me to explore it. Hemming Andersen, whose inventory of Danish scientific instruments was one of the first national inventories, gave me photocopies of Hauch's essay on the principles and decomposition of water, and of those pages of the Royal Society of London's *Journal Book* that dealt with Hauch's reading of that paper to the Society. Anja Skaar Jacobsen visited Toronto and introduced me to Hauch's role in the introduction of the antiphlogistic theory into Denmark.

Parts of this essay first appeared in Levere, "Lavoisier's Gasometer and Others. Research, Control, and Dissemination," in Marco Beretta, ed., *Lavoisier in Perspective* (Munich: Deutsches Museum, 2005), 53–67, and in "The Hauch Cabinet, Chemical Apparatus and the Chemical Revolution," *Bulletin of the Scientific Instrument Society* 60 (March 1999): 11–15.

CHAPTER 7

The Public Space of Knowledge and the Public Sphere of Science

Marie Thébaud-Sorger

My aim in this chapter is firstly to draw attention to the materiality of performances which forged the conditions of knowledge sharing. Instead of envisaging shows and displays as tools of dissemination, orchestrating a one-way educational transfer to the audience,[1] I wish to overcome traditional separations and rather to approach performances in terms of dynamic interaction, where intellectual process is at the heart of the material devices of the exchange. Performances, shows, and displays are a familiar subject for historians of science.[2] To take one example, the emergence of natural philosophy was based on experimentation with spectacular effects revealing the properties of matter. We have learned how the social materiality around the air pump built its scientific legitimacy and the emergence of a new physical space of knowledge that marked the birth of the laboratory, shaping the legitimacy of the people conducting the experiment as well as those admitted to testify to results.[3] Devices and instruments, conceived as mediation tools between nature, practitioner, and audience, reorganised the social world of science, and brought the material evidence that the investigation of nature required. According to Thomas Hankins and Robert Silverman in *Instruments and the Imagination,*[4] the interplay between actors and objects reveals three different levels: the first is defined by the instrument itself, the second considers the effect of the devices, and the last is the evidence and scientific theory revealed

---

1   Roger Cooter and Stephen Pumfrey, "Separate Spheres and Public Places: Reflections on the History of Science Popularization and Science in Popular Culture," *History of Science* 32 (1994): 237–267.

2   Bernadette Bensaude-Vincent and Christine Blondel, eds., *Science and Spectacle in the European Enlightenment* (Aldershot: Ashgate, 2008).

3   Steven Shapin, "The House of Experiment in Seventeenth Century England," *Isis* 79 (1988): 373–404; Steven Shapin and Simon Schaffer, *Le léviathan et la pompe à air* (Paris: La découverte, 1993), Collection <<Texte à l'appui>>.

4   Thomas Hankins and Robert Silverman, *Instruments and the Imagination* (Princeton: Princeton University Press, 1999).

© MARIE THÉBAUD-SORGER, 2022 | DOI:10.1163/9789004501225_008

by the mediation. They did not exclude instruments of pleasure, imagination and natural magic, stressing the fact that "the experiment and the amusement were not distinguishable." The interplay occurs between curiosity, credulity, and wonder. True objects of science appeared alongside amusing illusions; on the frontier of experimentation, we see the development of science as spectacle, attested from the whole of the early modern period and then especially in the eighteenth century by a growing audience eager for entertainment and whose inquisitiveness was generated by the wonder of display. This has given rise to a new historiography reconsidering the experiments of electricity, fireworks and ballooning for instance.[5]

By focusing on science in public spaces, I aim to explore more deeply "sites of science" that have been studied across various social and global contexts—in and beyond the scientific world—by looking more precisely at the material space in which, and for which, various public experiments were deployed at the city scale.[6] These reflected at several levels their geographic and cultural contexts. The rise of public science presents a convergent phenomenon at the European level in the early-modern period and the Enlightenment, whereas public displays of sciences echo many local variations where political, cultural, religious and economic features ruled the relationship between the different social spaces of sciences – such as the role of academic institutions, royal patronages, culture of amateurs, entrepreneurial developments. I wish therefore to examine, how the materialization of cutting-edge knowledge performed in various public places in France in the 1770–1780s, fostered the rise of an informed audience. Looking at public spaces takes in the collective and political dimension of public life, where the public should not be seen in the abstract but as a concrete gathering of individuals: the many "practitioners"

5   Simon Schaffer, "The Consuming Flame: Electrical Showmen and Public Spectacle in Eighteenth-Century England," in J. Brewer and R. Porter, eds., *Consumption and the World of Gods* (London: Routledge, 1993), 489–526.
     P. Bertucci & G. Pancaldi, eds., *Electric Bodies. Episode in the History of Medical Electricity* (Bologna: CIS, University of Bologna, 2001); Simon Werrett, *Fireworks: Pyrotechnic Arts and Sciences in European History* (Chicago: University of Chicago Press, 2010); Marie Thébaud-Sorger, *L'aérostation au temps des Lumières* (Rennes: Presses Universitaires de Rennes, 2009).
6   Stéphane Van Damme and Antonella Romano, *Science and World-Cities in the 16th and 18th Centuries*, special issue of the *Revue d'Histoire Moderne et Contemporaine*, 55 (2008); Lissa L. Roberts, "Geographies of Steam: Mapping the Entrepreneurial Activities of Steam Engineers in France During the Second Half of the Eighteenth Century," *History and Technology* 27(2011): 417–439; Bert De Munck, Antonella Romano, eds., *Civic Epistemology. Urban Knowledge in the Early Modern Period* (Leiden: Brill, 2019).

of the urban life:[7] reader, walker, spectator, consumer. My analysis is not restricted to what is shown but rather to the way in which it is shown and received: the complex course and process of the interweaving of empirical and concrete interactions and intellections. It is precisely the kind of mediation that occurred within interplays between audiences and objects that I will investigate, echoing socio-anthropological approaches that have studied demonstration processes, their peculiar forms, temporalities and social and political configurations.[8] This means exploring the way inventions, objects and devices were shown and, especially, which frameworks, locations and types of audience have played a crucial part of the making of knowledge. It is necessary to question the boundaries between science and technology and consider these shows as fully part of modern science as well as of a growing consumption culture, which emerged in 18th-century Europe.

The rise of public expertise has also questioned the relationship between science and society. The marketplace sketched out by natural philosophy enhanced the claim of useful purposes through the understanding and the mastering of matter that together link demonstrations from laboratory to workshop and manufactures.[9] And yet, little attention has been paid to mechanical inventions, although they fostered many curious public experiments, especially in cities such as London and Paris. As Liliane Hilaire-Pérez and I have shown elsewhere, exhibitions and experiments of inventive goods and devices occurred at several stages of their production and commercialization, involving different audiences (consumers, supporters, funders, administrators, city counsellors, learned societies). This was strongly linked to inventors' strategies, contingent on legal frameworks protecting inventions.[10] Instead of considering audiences as passive consumers, we showed how, between printed leaflets, advertisements, and the various places (workshops, sites) located in urban space, the public was called upon to play an active role in judging innovative

---

7   Michel de Certeau, Luce Giard, *L'invention du quotidien*. Tome 1. *Arts de faire* (Paris: Gallimard, 1990).

8   Marres Noorte and Javier Lezaun, "Materials and Devices of the Public: An Introduction," *Economy and Society* 40 (2011): 489–509; Claude Rosental, "Towards a Sociology of Public Demonstration," *Sociological Theory* 31 (2013): 343–365; Idem, *La société de la démonstration* (Vulaines-sur-Seine: Éditions du Croquant, 2019).

9   Larry Stewart, "The Laboratory, the Workshop, and the Theater of Experiment," in Bensaude-Vincent and Blondel, *Science and Spectacle*, 11–24.

10   Liliane Hilaire-Pérez and Marie Thébaud-Sorger, "Les techniques dans l'espace public. Publicité des inventions et littérature d'usage au XVIIIe siècle (France, Angleterre)," *Revue de Synthèse* 127 (2006): 393–428.

devices and projects. The performance and exhibition of various technical objects is certainly one of the places in which the transformation from curiosity to utility occurred, and where interest in useful knowledge formed beyond a culture of amusement based not only on spectacle, but also one of the spaces in which pioneering knowledge could be forged.

To tackle those issues, we need to scrutinize – as far as a reconstruction of the reception of performances is possible – the interplay between publics and objects embodied at a very concrete level: how "things" were staged and may have been diversely "perceived." In the first part of my paper, I focus on this new urban consumption culture which surrounded technical inventions, sustaining opportunities for demonstrations, spectacles and shows. I consequently address the issue of publicity and the scale of experiments. Focusing more specifically on "open air" experiments, I then highlight the way in which public reception changed as the objects' prominence in the public space increased. This was a long-run process for many devices. Case studies of ballooning in France between 1783 and 1784 will allow me to explore further these dimensions, and draw conclusions on the role these kinds of public experiment played in the appropriation of the new pneumatic chemistry.

1   Technological Displays: The Interaction Process and Publics as Experts

1.1   *Talking Heads Versus Balloons, Paris, Autumn 1783*

Whereas many historical studies have highlighted London as the main city for mechanical displays, a great interest for science and technical innovations dominated 1780s Paris.[11] In the wake of the phenomena presented by the beginning of the first balloon flights, the essayist Antoine de Rivarol (1753–1801) took the opportunity to published anonymously a witty and brilliant *Lettre à Monsieur le Président de *** sur le globe aérostatique, les têtes parlantes et sur l'état de l'opinion publique à Paris,*[12] where he depicted this Parisian craze for inventions. More well known for his conservative writings during the Revolution, this leaflet was written at the beginning of his literary career. Despite its ironic

---

11   "Culture matérielle, commerce et consommation de science à Paris," in Stéphane Van Damme, *Paris, Capital philosophique: de la Fronde à la Révolution* (Paris: Odile Jacob, 2005), 125–143.

12   *Lettre à Monsieur le Président de *** sur le globe aérostatique, les têtes parlantes et sur l'état de l'opinion publique à Paris,* 20 septembre 1783, Sylvain Menant, ed. (Paris : Les éditions Desjonquière, 1998). *Journal de Paris* (1 mai 1778), *Journal de Paris* (6 July 1783), 187.

tone, the narrative reflects the peculiar atmosphere of the French capital, where a growing consumption culture fostered the appearance of new sites of displays – showrooms and shops – for novel goods. These sites, alongside traditional fairs and theatres, included new walks (promenade) such as the Palais-Royal and pleasure gardens (Tivoli, Vauxhall, *redoute chinoise*),[13] mixing polite audiences: aristocrats, grande and petite bourgeoisie. Rivarol described a set of experiments using pioneering knowledge in which innovative and intricate technical devices were displayed to their late eighteenth-century Parisian audiences. His immediate aim was to promote a "mechanical masterpiece", the display of l'Abbé Mical's "talking heads." But he built his narrative by contrasting it with what was in fact the main object of his letter, a French craze, "madness" or "contagious illness", for the first balloon ascents.

What are these inventions about? The *talking heads*, a curious attempt to recreate the human voice artificially, have attracted historical interest focusing on the recording device and system of language,[14] or on automaton. The display of these mechanical marvels, sometime across Europe, have been widely studied, such as Vaucanson' famous automatons,[15] the machineries of James Cox, Christopher Pinchbeck, or more recently the *androides* of the Swiss watchmaker Jacquet Droz, all exhibiting cutting-edge technology.[16] Mical's device,

---

13   Jonathan Conlin, "Vauxhall on the Boulevard: Pleasure Gardens in London and Paris, 1764–1784," *Urban History* 35 (2008): 24–47; Laurent Turcot, "La fonction de la promenade dans les récits de voyage à Paris," *Dix-huitième siècle* 39 (2007): 521–541; Liliane Hilaire-Pérez, "Les boutiques d'inventeurs à Londres et à Paris au XVIIIe siècle: jeux de l'enchantement et de la raison citoyenne," in N. Coquery, ed., *La boutique et la ville, commerces, commerçants, espaces et clientèles* (Tours: CEHVI, 2000), 171–189.

14   Alfred Chapuis and Edouard Gélis, eds., *Le monde des automates* (reprint, Genève: Slatkine,1984); Hankins and Silverman, "*Vox Mechanica:* the History of Speaking Machines," in *Instruments and the imagination*, 186–7; Jessica Riskin, "Eighteenth-Century Wetware," *Representations* 83 (2003): 97–125. Tanaka (Suzuki) Yuko, "Préservation of French-Speaking Automatons and Their Pronunciations in 18th Century France, Focusing on l'Abbé Mical's *Têtes Parlantes*, and A. Rivarol's Letter of 1783," *Aesthetics*, 18 (2014): 13–27.

15   André Doyon and Lucien Liaigre, eds., *Jacques Vaucanson, mécanicien de genie* (Paris: Presses Universitaires de France, 1966); Simon Schaffer, "Enlightened automata," in W. Clark, J. Golinski and S. Schaffer, eds., *The sciences in enlightened Europe* (Chicago: Chicago University Press, 1999), 126–165.

16   Marcia Pointon, "Dealer in Magic: James Cox's Jewelry Museum and the Economics of Luxurious Spectacle in Eighteenth Century London," in N. Marchi and D.W Goodwin, eds., *Economic Engagements with Art* (Durham and London: Duke University Press, 1999), 423–451; Roger Smith, "James Cox's Silver Swan," *Artefact* 4 (2016): 361–365; Liliane Hilaire Pérez, "Technology, Curiosity and Utility in France and in England in the Eighteenth Century" in Bensaude-Vincent and Blondel, eds., *Science and Spectacle*, 25–42.; Adelheid Voskuhl, *Androids in the Enlightenment: Mechanics, Artisans, and Cultures of the Self* (Chicago: University of Chicago Press, 2013).

with its two imposing and giant heads like huge statues, made visible its internal workings, the display of its mechanism integral to the automaton itself. It was painted gold and decorated with a frontispiece. When Mical started the engine, air came shooting through the membrane of the false teeth to produce phonemes. These were arranged to pronounce entire sentences, the two heads having a "conversation" with each other.[17] After years of trials, the Abbé Mical put on his first experiment in June 1783, attracting the interest of members of the Royal Academy of Sciences.[18] His timing could not have been worse as this was the summer that saw aerostatic experiments reach Paris. The public's gaze turned to the heavens and Mical's invention attracted scant regard.

The Montgolfier brothers, provincial manufacturers, developed the lighter-than-air process which allowed their machines to defy gravity. Their discovery as well as its reception have been extensively studied.[19] The first public experiment took place in the public square in Annonay on 4 June 1783, on the same day as a political meeting of the *États du Languedoc*. Testimonies of the experiment drew the attention of the French Academy of Sciences, and Etienne de Montgolfier came to Paris in order to repeat his experiment in front of the Academy and Louis XVI. In the meantime, the event aroused curiosity. Another undertaking in Paris was led by a well-known lecturer Jacques Alexandre César Charles who, with the support of other important figures of the aristocratic elite such as the Duc de Chartres, opened a subscription to carry out a similar attempt which took place 27 August at the Champs de Mars. Charles' experiment, however, used a different technical process from the Montgolfier brothers. He used flammable air (hydrogen), a gas that the Montgolfiers had wanted

---

17   Tanaka (Suzuki) Yuko, "Préservation of French-speaking automatons."

18   Faujas de St Fond wrote twice to Franklin to bring him over to attend the experiment and be invited to have dinner with the inventor, *Letter to Benjamin Franklin from Barthélemy Faujas de Saint-Fond*, 16 June 1783 and 18 June 1783, *Founders Online,* National Archives, https://founders.archives.gov/documents/Franklin/01-40-02-0107: "J'ai vu et entendu moi-même ces deux têtes, qui n'ont encore été vues par personne; et je puis vous assurer que C'est un ouvrage Bien etonnant, qui n'est peut être pas encore à son dernier degré de perfection, mais dont la mécanique est faite sur les plus Scavants principes." Faujas would become also a key figure behind the promotion of aerostation.

19   James Martin Hunn, *The Balloon Craze in France: A Study of Popular Culture, 1783–1799* (Nashville: Vanderbildt University, 1982); Charles Coulston Gillispie, *The Montgolfier Brothers and the Invention of Aviation, 1783–1784* (Princeton: Princeton University Press, 1983); Marie Thébaud-Sorger, *'L'air du temps.' L'aérostation: savoirs et pratiques à la fin du XVIIIe siècle, 1783–1785* (Doctoral thesis, EHESS, 2004; published as *L'aérostation au temps des Lumières*). The historiography has been growing on this topic since. See Michael R. Lynn, *The Sublime Invention: Ballooning in Europe, 1783–1820* (London: Pickering and Chatto, 2010); Mi Gyung Kim, *The Imagined Empire: Balloon Enlightenments in Revolutionary Europe* (Pittsburgh: University of Pittsburg Press, 2016).

to use but, encountering too many practical difficulties, substituted with "air rarefied by heat" (hot air), which was also considered to be a kind of "gas." Some days later Montgolfier reproduced their experiment of a hot air balloon in front of the royal court in Versailles.[20] The two flights resulted in a competition, sparking intense exchanges in newspapers and among enthusiasts over the processes, even before scientific authorities were able to pass judgement.

By recreating voices able to pronounce whole sentences or conferring on humans the power to escape gravity, both inventions appeared to be at the frontier of the impossible. When Rivarol wrote his letter in September 1783, balloons were still unmanned. Yet in just over a month the first human flight would take place and, with it, wild expectation over the prospect of unbounded travel and adventure. Rivarol sought to deflate those expectations by stressing the hollowness of the ballooning discovery, a fashionable invention arousing emotional extremes and ridiculous behaviour. The "talking heads", in contrast, attracted a "polite", "learned", and conversant audience. Rivarol employed the classic antidemocratic criticism of the volatility of public opinion in undermining the merits of ballooning, claiming that this invention should be endorsed only depending of its utility. Conversely, the talking heads, the fruit of "genius" and craftsmanship, were presented as a work of art which, in recording speech, could result in a repository of the spoken word, a veritable oral archive. Yet, the talking heads and balloons relied more on the sensation provoked by prodigious effects than perceived as immediately useful. Taking into account how the material conditions of their exhibition have affected their audiences reveals the persistent interest in technical prowess and ingenious processes, beyond any consideration for their alleged utility. Despite his disdain for the public spectacle of ballooning, Rivarol in fact validated the educational role played by the senses and emotions, and acknowledged one useful outcome of balloon ascents, which was "to have freed man from this unassailable law of gravitation which constantly brings him back to the surface of the earth."[21] According to Rivarol, competition between novel inventions had less to do impartial judgement on their intrinsic value and potential than on the way they were displayed and staged for the wider public.

The relevance of his testimony lies less in his narrative, with its strong bias and rhetorical flourishes, than in the manner in which he depicts theses different audiences: learned experts, and an elegant, genteel, and poised elite contrasted with a common public of the streets. Whereas public displays were

---

20   Catherine J. Lewis Theobald, "Soaring Imaginations: The First Montgolfier Ballooning Spectacle at Versailles in Word and Image," *Lumen* 39 (2020): 23–53.

21   Rivarol, *Lettre à Monsieur*, 172–173.

often interpreted as testament to the openness of knowledge and a democratization of experiencing the progress of human understanding, they were also overlaid with social prejudice. The proliferation of public displays and the growth of audiences, whose senses and feelings of awe were engaged in witnessing and appraising experiments, could have also presented a pitfall whereby entrepreneurs and inventors were suspected of trickery. Credulous publics could be deceived through slights of hand, or a crafty and mysterious setting – such as Kempelen's famous chess player or, as Rivarol communicated in a note, the ridiculous fake "talking doll" displayed on the boulevard. While it was easy to discredit magic tricks, it was much harder to cast doubt on the testimony of a large crowd in the centre of Paris witnessing the ascent of a balloon. Large mixed audiences, which included the uneducated and the ignorant, were frequently depicted as naïve, mesmerized, and transported by the latest craze. Any experiment that was greeted with great and sustained enthusiasm was open to suspicion by this very fact, and its audience disqualified from making an enlightened judgement.

Rivarol also mentioned an experiment by the apothecary Antoine Quinquet (1745–1803) who succeed through an electrical machine to convert water vapour into frost, artificial hail, and even snow. But because Quinquet's demonstrations remained in his physic cabinet they "did not cause the same sensation as the balloons, because in order to seize the people's attention, hail, or snow would have to fall on command in the middle of the public square."[22] Similarly, the failure of Mical's experiment, according to Rivarol, lay precisely in the fact that its audience was too restricted, an interesting paradox given his disdain for balloon audiences. Even if members of the social and scientific elite were present at demonstrations, this was no guarantee of success. Public opinion, forged through exhibitions and displays, encompassed a world that extended well beyond polite society and academic appraisal. Thus, opinion played a major role in the market for invention and furthered the development of innovative devices. The Abbé Mical eventually took steps to make his show more visible or, in Rivarol's words, displayed for the "public curiosity", moving his workshop to a bigger and more visible venue where "the eyes" and "ears" of the public would experience his work in a setting conceived for its own performative function. The complex constellation of technical invention or experimentation encompasses the performative function of the physical venue, and a physical audience with all its senses engaged. Considered as such, a gathering

---

22   Ibid, 181.

of thinking individuals, as opposed to an informal and distracted crowd, which attest to the validity of the technical display or experimental process raises numerous questions about the transformation of the crowd, the mixed and nebulous audience into a public sphere.

1.2   A Flourishing Consumption Culture of Science and Techniques

The progressive involvement of the "public at large" in displays and exhibitions resulted in a change in the culture of recognition procedures. The Royal Academy of Sciences enforced its power upon technical inventions during the second half of the century, reviewed inventors' work alongside providing expertise for the bureau du commerce.[23] Mical was not a projector or a skilled entrepreneurial artisan, but a clergyman and science enthusiast seeking recognition and, of course, the approval of the academy, following the usual path of accreditations. Rivarol overstated the Academy's "approval" of Mical's invention while at the same time downplaying the Academy's support for Montgolfier. The Academy's on Mical's talking heads was very different from Rivarol's embellished description. It was positive, but hardly enthusiastic, finding fault with the voice's modulation, which made it sound "supernatural" and difficult to understand, as words were often inarticulate. As for Montgolfiers' balloon, the Academy was actually heavily involved. During the summer it had created a special interdisciplinary commission to assess the Montgolfier process (extending its remit to include Charles's trials), while the support of the king, who reimbursed the Montgolfiers for their new experiments in Paris, was clear. The popular interest in the first aerostatic flights affected the way scientific bodies and political institutions supported the invention (turning ballooning into a symbol of French genius).

A flourishing of science and techniques consumption culture fostered the emergence of the public as participant which then became a referee of inventions. Demonstrations of technical achievement were not indeed restricted to validation systems – different from country to country – and in France obtaining a privilege was not enough to convince of the relevance of an invention, thus the inventor would carry out demonstrations in order to consolidate his customer networks. Even if demonstrations were not performed in front of large audiences, each aimed at building spheres of legitimation. Inventors in the public sphere also built their markets and reputations by advertising in the press. Even if peculiar inventions such as Montgolfiers' and Mical's were not

---

23   Liliane Hilaire-Pérez, *L'invention technique au siècle des Lumières* (Paris: Albin Michel, 2000).

designed to be replicated or sold, they did not escape the logic of the market. Mical eventually sought to sell his mechanism to the Crown, principally as a way to recover his costs. When he eventually opened up his demonstration to a wider audience, the high entrance fee (3 livres) restricted his audience.[24] Behind ballooning undertakings, various entrepreneurial networks have been highlighted, for instance sulfuric acid factories for gas balloons, or the Parisian paint paper factory of Jean Baptiste Réveillon where Étienne Montgolfier sat up his workshop. He worked there with many collaborators, such as Amy Argand, an inventor and entrepreneur of a new oil lamp. For the brothers Montgolfier, who wished to obtain the rewarding title of "royal manufacture", seeking new outlets for their paper factory remained a strong motive. To foster relationships with an aristocratic clientele and wealthy bourgeoisie, who were prominent in circles of power, entrepreneurial networks became critical to supporting and promoting technical demonstrations, turning the elite attraction for curiosities into appealing and useful knowledge.

Travel accounts also reflected this appetite for novelties,[25] such as the narrative of the English traveller Anna Cradock, or of the Breton De Rouaud, who described their walks in Paris and its surroundings.[26] Both visited wallpaper factories (those of Arthur and of Réveillon),[27] the St Laurent Fair, and Marly's new hydraulic machine,[28] a well-known curiosity. Ms. Cradock was particularly interested in manufacturing processes, and she attended all of Paris's aerostatic experiments for 1784.

As Liliane Hilaire-Pérez has underlined, such strategies linking the pleasure of eyes and curiosity, notably for mechanical arts developed by craftsmen, skilled clockmakers and mechanics, were widely used by English and French inventors alike.[29] Newspapers, leaflets, advertisements and calls for

---

24 *Journal de Paris,* 92, April 1784, 409. He exhibited the machine twice a day at "noon" and "17 o'clock."

25 Turcot, "La fonction de la promenade."

26 Anna Francesca Cradock, *Journal de Mrs Cradock, voyage en France (1783–1786)* (Paris : Perrin, 1896); De Rouaud, *Voyage de Paris en 1782, journal d'un gentilhomme breton* (Vanne: Lafoyle, 1900).

27 Christine Velut, "Les stratégies de promotion de 'l'invention ordinaire' dans un groupe professionnel en formation: les fabriques de papiers peints à Paris dans la deuxième moitié du XVIIIe siècle," in M.-S. Corsy, C. Douyère-Demeulenaere and L. Hilaire-Pérez, eds., *Les archives de l'invention. Écrits, objets et images de l'activité inventive* (Toulouse: Université Toulouse le Mirail: collection Méridiennes, série Histoire et techniques, 2006), 559–571.

28 Thomas Brandstetter, "La machine de Marly: un spectacle technologique," *Bulletin du Centre de recherche du château de Versailles,* [online] 2012. https://doi.org/10.4000/crcv.11907

29 Liliane Hilaire-Pérez, "Technology, Curiosity and Utility."

subscriptions reveal this growing business and the diversity of what was on offer. For instance, the *Avant coureur* of the 1760s,[30] and in the next decade the various general periodicals such as the *Journal de Paris* (the only daily paper starting from 1777), reported on experiments, published announcements of novelties, recorded competitions and public events. It was therefore common to combine collective practices of reading with the attendance to various events *in situ* like displays, experiments, and lectures. De Rouaud described for instance that in 12 July 1772, he went to Ménilmontant to

> visit the house of M. Detienne, former infantry captain, knight of St Louis who found the secret of a putty which he covers the roof tiles with, without hiding them, and which makes them completely waterproof [...]. This house is admired by *tout Paris* and the court. M. Detienne promised the king that in two months he would reveal his secret to the public [...]. We then went to see automatons imitating human figures.[31]

The *Journal de Paris* made the announcement for the D'Étienne exhibiting house ten days earlier, specifying that "People who are interested in this discovery can come to M. D'Etienne, who will be pleased to receive them and satisfy their curiosity,"[32] the latter explaining to the visitors its revolutionary tiling system that kept the entire house water-tight, stunning views of Paris, and roof terrace with a complex system of fountains feeding an ornamental garden.[33]

The public sphere and its readership should not at this stage be separated from the actual audiences that attended the great variety of shows and demonstrations, fostering the emergence of new spaces of mixed sociability such as public classes, "*musées*", and societies founded by subscription.[34] Public

---

30   *L'Avant-Coureur, feuille hebdomadaire où sont annoncés les objets particuliers des sciences et des arts, le cours et les nouveautés des spectacles, et les livres nouveaux en tout genre,* published between 1760 and 1773.

31   De Rouand, *Voyage de Paris en 1782*, 37.

32   *Journal de Paris* 182, July 1st 1782, 746.

33   Jean D'Étienne, *Mémoire sur la découverte d'un ciment imperméable à l'eau et sur l'application de ce même ciment à une terrasse de la maison de l'auteur* (Paris: l'auteur, 1782); Valérie Nègre, *L'art et la matière. Les artisans, les architectes et la technique (1770–1830)* (Paris: Classique Garnier, 2016), 149–151.

34   Hervé Guénot, "Musées et lycées parisiens (1780–1830)," *Dix-huitième siècle* 18 (1986): 249–267; Daniel Rabreau and Bruno Tollon, eds., *Le progrès des arts réunis, 1763–1815 :* Mythe *culturel, des origines de la Révolution à la fin de l'Empire?*; Actes du colloque international d'histoire de l'art, Bordeaux-Toulouse, 22–26 mai 1989 (Talence: CERCAM, 1992); Michael R. Lynn, "Enlightenment in the Public Sphere: The Musée de Monsieur and Scientific Culture in Late Eighteenth-Century Paris," *Eighteenth-Century Studies* 32 (1999): 463–476.

assemblies were open and women were welcomed. At the public assembly of the Abbé Beaudau's *Société d'émulation* – which launched competitions encouraging a wide participation open to ingenious artisans – machine parts and models were exhibited. In June 1779 for instance, award winning devices were described in a report read by the amateur of sciences and promoter of techniques Dufourny de Villiers who also, according to the *Journal de Paris*, gave "demonstrations" of these novel devices.[35] At the *Musée de Monsieur*, Jean-François Pilâtre de Rozier performed sophisticated and striking experiments on inflammable air. These served as prelude to the promotion of an *antimephitic mask* he invented to protect cesspool workers from asphyxiation; he later undertook *in situ* demonstrations of the mask, but this time for an audience of academic experts.[36]

Beyond a narrow scientific and academic group, a large community of "improvers" undertook trials, demonstrations, and experiments for varied audiences within the city: workshops, parks and riverbanks all became sites of invention and display.[37] Dufourny, for instance required from the House of the King department (which help many inventive experiments), the right of access to the Pont Neuf to make a set of trials "to observe the action of deadwood on the surface of the water."[38] Technological displays became a common process for entrepreneurs, craftsmen and inventors, for all faced the same challenges: to build a market; gain recognition; protect their invention in the subtle balance between openness and secrecy.

Far from being restricted to goods that were "ready to sell", most technical processes and objects were subject to change. Many entrepreneurs endeavoured to promote new methods or inventions by calling on the public to participate in development and experiments. The rise of public expertise sometimes even overtook scientific- or improvement-society assessments. Publics were invited to share their own judgment on innovative devices, materials, commodities, and processes, thereby sustaining an emancipatory claim that combined economy with general improvement. Public curiosity sustained, on one hand, the

---

35   *Journal de Paris,* 30 June 1779, 739.
36   Marie Thébaud-Sorger, "Le musée scientifique autour de 1785. Entre curiosité et utilité: Les usages d'un lieu," in V. Milliot, P.Minard and M. Porret, eds., *La grande chevauchée : Faire de l'histoire avec Daniel Roche* (Genève: Droz, 2011), 449–462. See also the example of Scanégatty demonstration in Rouen in, Idem, "Capturing the invisible: Heat, Steam and Gases in France and Great Britain. 1750–1800" in Lissa Roberts and Simon Werrett, eds., *Compound History: Materials, Governance and Production, 1760–1840* (Leiden: Brill, 2017), 85–105.
37   Liliane Hilaire-Pérez, *L'invention technique,* 232.
38   Archives nationales de France (ANF), Maison du Roi, O1/1293/39.

possibility of financial support and moral credit (through patrons, subscribers, and consumers) for which an understanding of the process at stake was crucial. On the other, one can discern the outline of an emerging public sphere in which understanding and expectation of technical devices nurtured imagination, speculation, and even the possibility of social and political reform. The emergence of new concerns regarding the common management of everyday city-life, such as travel-safety and comfort, or the improvement of public spaces, gave birth to numerous small inventions and demonstrations involving several types of audience in the social hierarchy, including town aldermen, social notables, fellows of the academy, and an aristocratic elite.

1.3   *Mapping Demonstrations and the "Emancipated Spectator"*

As we have seen, the great diversity of the technical displays in an urban setting could not be reduced to a unique category nor condensed to performances in a physical place;[39] they could have encompassed multiples socio-material shapes, according to their locations (indoor/outdoor), and their different features (static/dynamic). The necessity to lead demonstration depends also on their own chronology in the validation processes of the market, upstream (for experts, subscribers, patrons) or downstream (possible consumers). Furthermore, between the testing of small-scale models, the demonstration of instruments and machines, or guided visits to a building in a city, they engaged spectators' senses as well as shaped the action of the audience. The question of audience composition remains prominent as is made clear by the popularity of some spectacular devices. This public was not always the "polite and learned" audience, nor members of various institutions, nor educated members of new restrictive sociabilities or public classes, nor even the credulous consumers of new goods, promoted by entrepreneurs outbidding their expected efficiency.

These different categories of publics were mainly forged by the actors of the time, part of a larger moral and philosophical debate about sensitivity and theatrical display. Their manipulation as rhetorical features serve mostly to value or discredit demonstrations rather than reflecting accurately reception processes that were actually at work. Therefore, historians need look also at other sources such as individuals' travel accounts, diaries, and letters. Certainly emanating from educated people, and despite being partial, they allow us nonetheless to shed another light on what was at play in the interaction, emphasizing mostly the intelligibility of what were displayed was actually never separated from its formal expression.

---

39   Rosental, "Towards a Sociology of Public Demonstration."

On one hand, demonstrations should be seen as a succession of sequences. They were given often with explanations in an oral presentation by the inventor or demonstrator,[40] or could be supplemented with printed instructions, leaflets and engravings. The audience was able to connect these forms of information with each other and assemble different contextual segments (social expectations, competing experiments etc.). For instance, the meaning of specular experience of inflammable air performed by Pilatre de Rozier in his *musée* in front of a polite audience, which could appear as a curious experiment of chemistry, is also referring to the issue of fighting workers' asphyxia in the wider context of seeking to improve the salubrity of the city. In front of various machineries displayed, whether it is an automaton, or a new pump, audiences could have been led to draw comparison with other artefacts of the same kind already seen. On the other, audiences were physically engaged in various practices in an active way: walking, talking about commodities, enjoying a curious show, being called upon to evaluate, or drawn into pedagogical interaction.[41] Each individual could *a priori* make comparison and use of their critical spirit, at the same time as enjoying novel experiments. One should, therefore, challenge the "categorization" of audiences, by taking into account different parameters such as places, practices, scales, temporalities shaping the way interactions occurred.

Some public demonstrations performed as a single dynamic event, offered a specific drama, a sharing of feeling, or, in the words of Jacques Rancière, "*le partage du sensible*" [42] for individual spectators within a whole audience looking simultaneously at the same display. This is especially the case when demonstrations occurred as an "open air" experiment, although many "open air" trials were not all "public." Many experiments carried out by the Academy of Sciences in France, were conducted in precise conditions and to a restricted community of "experts" able to assess an experiment's success and potential. The park at Versailles provided an isolated space where tests were conducted

---

40   Bret Patrice, " Un bateleur de la science: le 'machiniste-physicien' François Bienvenu et la diffusion de Franklin et Lavoisier," *Annales historiques de la Révolution française*, 338 (2004): 95–127.

41   John R. Millburn, *Benjamin Martin, Author, Instrument-Maker, and "Country-Showman"* (Leiden: Noordhoff, 1976); "Science Lecturing in the Eighteenth Century," Special issue, *British Journal for the History of Science* 28 (1995), and the work of Desaguliers examined by Larry Stewart, *The Rise of Public Science* (Cambridge: Cambridge University Press, 1992).

42   Jacques Rancière, *Le spectateur émancipé* (Paris: La fabrique éditions, 2008); as *The Emancipated Spectator*, Gregory Elliott, trans., (London and New York: Verso, 2011).

before a select audience.[43] Official authorizations were required by, and granted to, inventors in Paris in order to avoid gatherings of passers-by who could interfere with the running of experiments. Locations and times were all cautiously considered. Large-scale experiments were sometimes necessary owing to the size of the mechanics. This implied further costs (usually shouldered by inventors) and organization, the commitment of the city, and even police supervision, especially in case of public experiment.

On December 3, 1777, for instance, on the Louis XV Square, Prospère Siméon Hardy noted in his diary that the police lieutenant Lenoir attended "the test that had to be made in order safeguard houses from the activity of fires." The staging exhibited small timber houses coated with the fireproof putty. One of the houses was set on fire "in the presence of a large number of curious spectators who had been attracted by the spectacle."[44] The trial turned out to be unsuccessful. Hardy reported booing from the audience, no doubt disappointed by the fact that their expectations had not been met. Their disappointment however didn't rely only in the failure of a promising spectacular show but rather, according to Hardy, because they were expecting the reproduction of an English process they have heard about. The audience seemed therefore relatively well informed. Such spectacular performances, and the reactions to them, raise numerous questions: how do we approach the kind of scientific and technical knowledge involved and shared? What were the interplays between individual judgment and collective expectations? Why did it matter to perform these experiments in public setting – and especially those regarding the improvement of social and urban life?

The shift from remote unseen confined places to larger urban venues and public spaces involved authorizations, police management or control, and a guarantee that, beyond polite audiences and institutional bodies, a wider audience should be involved at one stage or another. But the scale of the experiment as well as its publicity defined conditions that tended to regulate its legitimacy. On these thin boundaries, we should consider that open-air experiments could be excessively different depending on their scale and their publicity. With an experiment in the middle of the city, the crowd could be transformed into a common public whose testimony transformed the technology into a question of the common good and citizenship. This was part of the knowledge production and dissemination.

---

43 For instance, see a certain Tremel experiment on models of ships on the Grand canal: ANF, Paris, 138AP/212 fonds Daru, Letter from Jean-Claude Pingeron to Pierre Dufourny de Villiers, 12 February 1784.

44 Bibliothèque nationale de France, ms 6682, fol 422, *Mémoires de Pierre-Siméon Hardy*, 3 décembre 1777.

2   The Ballooning Workshop

2.1   *Preparing Flight: A Collective Enterprise*

In this section my aim is to give deeper insight into the role of open-air experiments in knowledge making. This will allow us to identify this kind of interplay as a major "space of knowledge." The first balloon ascents provide a wonderful case study because the enthusiasm they raised multiplied the opportunities, the various reflections, and the dissemination of a new technology based on an understanding of aerial fluids and the identification of different airs and gases. Aerostation issues played a large part in the circulation within and beyond a group of chemists who practised cutting-edge analysis of combustion processes (which were actually very complex to conceptualize) as well as flammable air (hydrogen) whose very nature provided the basis of many assumptions.[45] Moreover, rather than a few demonstrations, which was usually the case for expensive technical inventions carried out on a large scale such as pumps, fire engines and fire extinguishers, navigation improvements, etc.), balloon ascents offered a series of experiments that attracted huge crowds, creating an impressive market of prints, engravings and objects. Becoming a social and cultural prominent event not only at the end of the French *ancien régime*, but more generally in European culture, they have left many traces in archives (private letters, diaries) as well as in state correspondence and police management. The flight of Charles and Robert, for instance, from the Jardin des Tuileries on I December 1783 left witnesses spellbound, generation more than a hundred engravings (including caricatures) and numerous reports in private and public papers. On one hand, this allows us to recognize this specific event as a "craze" or even "madness",[46] as Rivarol noted and, on the other, makes feasible a possible reconstruction of the various responses through comparing testimonies and various reports and prints.

Before the spread of this technology throughout England and in other European countries, my survey of the first balloon attempts reveals that no fewer than 20 human flights took place in France in 1784. These flights were led by a group of heterogeneous practitioners, lecturers, clergymen, merchants,

---

45   Marco Beretta, *The Definition of Chemistry from Agricola to Lavoisier.* Uppsala Studies in History of Science, 15 (Canton, MA: Science history publication, 1993); Maurice Crosland, "'Slippery substances.' Some Practical and Conceptual Problems in the Understanding of Gases in the Pre-Lavoisian Era," in F. L. Holmes and T.H. Levere, eds., *Instruments and Experimentation in the History of Chemistry* (Cambridge, Mass.: The MIT Press, 2000), 79–89.

46   Hunn, *The Balloon Craze in France.*

instrument makers, architects and enthusiastic amateurs.[47] The number of small-unmanned balloons launched was so great that regulations had to be drafted in order to forbid such practices.[48] The social organisation of those experiments, which took place in French provincial towns show common characteristics. Focusing specifically on the Nantes case, we can discern a number of important features: the type of enterprise, the support of local powers, the creation of the public sphere and consensus around progress and techniques while fostering the circulation of pioneering knowledge on pneumatic chemistry including, but going far beyond, the research carried out in the leading laboratories.

Balloon flights have long been considered as "entertainment for the masses" or "spectacular performance,"[49] but debates into the new chemistry of gases, and the nature and properties of "airs" were at the heart of the invention. This chemical framework has been examined but rather as a secondary factor. Recent approaches have also outlined how Lavoisier's experiments into the decomposition of water coincided with research into gas production for ballooning purposes (until the French Revolution).[50] Thus balloon ascents may have been seen as an opportunity to advance the development of this new chemistry. Yet, it seems that each attempt performed in the middle of the public space was, in fact, more an expression of the complexity of the intellectual landscape around the properties and definition of "airs." Authorizations obtained for flights were also the outcome of a complex interplay between local or city regulations, administration and scientific expertise provided by local academies in order to obtain permission for the flights.[51]

---

47 Richard Gillespie, "Ballooning in France and Britain, 1783–1786: Aerostation and Adventurism," *Isis* 75 (1984): 249–268.

48 Marie Thébaud-Sorger, "Innovation and Risk Management in Late Eighteenth-Century France: The Administration of Inventions in French Cities at the end of the Ancien Régime," in Christelle Rabier, ed., *Fields of Expertise. A Comparative History of Expert Procedures in Paris and London. 1600 to Present* (Cambridge: Cambridge Scholar Press, 2007), 261–189

49 Simon Schaffer, "Natural Philosophy and Public Spectacle in the Eighteenth Century," *History of Science* 21 (1983): 1–43; Barbara Maria Benedict, *Curiosity. A Cultural History of Early Modern Inquiry* (Chicago: University of Chicago Press, 2001; Paul Keen, "The 'Balloonomania': Science and Spectacle in 1780s England," *Eighteenth-Century Studies* 39 (2006): 507–535.

50 Janis Langins "Hydrogen Production for Ballooning During the French Revolution: An Example of Chemical Process Development," *Annals of Science* 40 (1983): 531–558. Thébaud-Sorger, Op.cit. (note 32); Mi Gyung Kim,"'Public' Science: Hydrogen Balloons and Lavoisier's Decomposition of Water," *Annals of Science* 63 (2006), 291–318.

51 Marie Thébaud-Sorger, "Innovation and Risk Management."

The apparent impossibility of a heavy machine ascending against the force of gravity piqued a public's interest. The difference between types of airs provided the necessary framework to turn this marvel into an acceptable, rational, fact.[52] By giving a comprehensible expression to a phenomenon that had hitherto been thought impossible, one can consider the reproduction of open-air flights as an experimental fact, wherein control over construction as well as the competence of the actors became more or less collectively established. The instant at which the balloon took flight occurred as a drama, forged by a unity of time, space and action, before the eyes of thousands of people. The aerostatic staging presented a theatre that transformed the experiment into a dramatic story through a collective, urban celebration. As reported in an account of the first experiment in Nantes on 14 June 1784: "This spectacle was, perhaps rightly, regarded as one of the most beautiful ever seen in this city."[53] The whole city joined in, and the description of a truly enormous audience is reflected in each report. The *Courier de l'Europe*, in describing the second ascent on 6 September 1784, reported "at least a hundred thousand souls watched the launch of the *Suffren*."[54] The gathering is variously defined as a "crowd", "a prodigious concourse of audiences", "a huge number of people", or "citizens from all ranks and social orders."

Provincial ballooning entrepreneurs ranged widely over a heterogeneous group of professions, including teachers, artisans, manufacturers, and scientists. To this group we have to add amateur enthusiasts who came together and helped to carry out various projects[55] by raising monies, obtaining materials, and authorizations. In this, these individuals gained symbolic reward – and social and scientific recognition – none of them demanded money, as entrepreneurs of science usually did. These enthusiasts synthesized of many different skills whether social, or scientific (mathematics, chemistry) and practical know-how (tinplate making, sewing, varnishing and painting, for instance). In Nantes, two undertakings converged to produce the city's balloon, *Le Suffren*: a group of merchants and aristocrats such as Coustard de Massy, Knight of Saint Louis the father Mouchet, an Oratorian professor of physics, and another group of amateurs led by Pierre Leveque, professor of hydrography who possessed no

---

52   Lorraine Daston, "Marvelous Facts and Miraculous Evidence in Early Modern Europe," *Critical Inquiry* 18 (1991): 93–124.

53   *Supplément aux affiches de Bretagne*, n°25, pp.227–228: Procès-verbal de l'expérience aérostatique faite à Nantes, le 14 juin 1784, Nantes 1784.

54   *Courrier de l'Europe*, Juillet 1784.

55   Richard Gillespie, "Ballooning in France and Britain."

relevant chemical knowledge,[56] but who worked with the help of Louvrier, an apothecary who conducted chemistry lessons in Nantes.[57]

Explanations of the technical and scientific choices for the balloons made by local enterprises were advertised in newspapers, during public courses, or on the occasion of visits to the workshops. This provided the means of keeping the project alive and interesting, while building up confidence in a venture which was hazardous. The phenomenon opened up a true experimental space. The Nantes undertaking aimed to reproduce a gas balloon. Its backers embarked on research in order to master the different steps of the chemical reaction employing sulphuric acid mixed with water poured over iron shavings. They scrutinized and compared the fineness and quality of shavings from several local factories, and experimented with different proportions of dilution of acid with water in several barrels. The latter were kept as evidence at the home of the town councillor Guérin de Beaumont.[58] The silk taffetas were also varnished at M. Guyot et Diet's factory. Tin-plate makers were also often required to produce specific devices like valves, pipes and cooling systems. Limiting the cost of gas production was at the heart of many a debate. The idea was to find a cheaper alternative to the sulphuric acid method, or to control the processes of producing inflammable air more effectively by avoiding warming which occurred during the chemical exothermic reaction of acid poured on metals (with several items of apparatus and the relevant skill), or by using the right proportion of water and acid. While in Dijon, the chemist Guyton de Morveau, who led the flight, saw an opportunity for developing his own research and proposed an innovative "potato gas." The experiment was not successful.[59] In Nantes, Leveque, through his correspondence with the

56   Musée de l'Air, Le Bourget, Fonds Montgolfier, XIII-28, Lettre de Pierre Lévêque à Etienne de Montgolfier

57   Christine Nougaret, "Les débuts de l'aérostation 1783–1784," *Mémoires de la société d'histoire et d'archéologie de Bretagne* 61 (1984): 165–191, at p.169 and Anne-Claire Déré, "Les aérostats, une industrie de progrès ?" in *Un musée dans sa ville, science, industrie et société dans la région nantaise, XVIII-XX siècle*, Jean Dhombres, ed., (Nantes: Quest Editions, 1990), 98–110.

58   Archives Municipales, Nantes, FF 276, n°10 lettre du procureur syndic Guérin de Beaumont sur ces libelles injurieux; Réponse du rédacteur du PV de l'expérience aérostatique du 14 juin 1784 au père Budan, Nantes le 21 juin 1784, chez Brun.

59   Guyton de Morveau et al., *Description de l'aérostat 'l'académie de Dijon' contenant le détail des procédés. Suivi d'un essai pour une application de la découverte de Montgolfier à l'extraction des eaux des mines* (Dijon: Causse, 1784); Leslie Tomory, *Progressive Enlightenment. The Origins of the Gaslight Industry, 1780–1820*, (Cambridge, Mass.: MIT Press, 2012); Marie Thébaud-Sorger, "'Nation fière et 'nation légère'. L'Angleterre, la France et l'invention des ballons," *Documents pour l'Histoire des Techniques* 19 (2010): 229–241.

Academy of Sciences, applied a few ideas put forward by Meusnier de la Place, who worked with Lavoisier on the decomposition of water, applying tinplate boxes dedicated to washing and cooling the gas produced in the barrels. In this method, pipes of 3 inches in diameter and 5 feet in length joined the barrels and the tinplate boxes full of an alkaline solution in order to purify the gas of its carbonic acid.[60]

2.2   The Rise of the Balloon: An Urban Dramaturgy

The day of the launch many different groups were involved: restricted audiences, along with authorities who kept a close and careful eye on the enterprise, and an audience of readers and subscribers among a wider public, all attending the flight, forming a large crowd. But each could catch a glimpse of the processes at stake, thanks to complementary information contained in engravings (even cheap and popular ones) and journal reports advertising the project. What did the audiences see and what were they able to understand?[61] The apparatus exhibited was highly visible, as if on a theatrical stage, providing the subjects of further prints and narratives. The most innovative research was thus shown to a wide variety of people. By endeavoring to pay attention to the material conditions of the reception: such as, how the stage was organized on a very basic level eventually led to assumptions about the way these experiments could help in disseminating cutting-edge knowledge within a new visual and sensory culture. In any case, the wonder of the ascent could be separated from the understanding of the principle and techniques that had driven its construction. Even without being familiar with all the chemical and technical details of different experiments being conducted, and which were otherwise available via the prints sold by entrepreneurs after the flight, as in Nantes and Dijon, one can assume that each element visible on the stage constituted a piece of the narrative and the intelligibility of a process that could be easily and widely comprehended. These elements are fully part of the interest of the public, as it is revealed in a letter from the enthusiast Pierre Le Gouz, sent to his friend Mauduit in Nantes, asking to keep him aware of the controversy between the Oratoriens and Levêque – through leaflets printed during the summer and where the scientific arguments of each part is debated.[62] Many newspapers reports have lampooned also behaviours of the uncultivated audiences, but also of the aristocracy. Women were an especial target

---

60   *Description faite de la seconde expérience aérostatique faite à Nantes. le 6 septembre 1784, sous direction de M.Levêque, correspondant de l'académie des sciences* (Nantes, 1784).
61   Rancière, *Le spectateur émancipé*.
62   Archives Municipales, Nantes, série I I. 146. Lettre de Legouz à Mauduit, 31 août 1784.

for ridicule. They were depicted chatting confusedly about "inflammable air."[63] Where chemistry was the object of conversations, crucial arguments of the experiment served as the framing for discussions: the definition of "flammable air" became shared by very large part of the public. The public understood that knowledge of chemical substances was as essential in the understanding of flight. And this understanding and knowledge extended to all categories and social classes attending displays of flight, from idle rentiers, their domestics, children, to various liberal professions such as clerks, physicians, architects, and many artisans, merchants and their wives, artists, writers. Enthusiasm for ballooning was such that many shops closed and workers took their day-off to attend a flight. These behaviours provoked debates within local administration about the value of such displays to the local economy. Yet it was a great urban celebration and everyone's presence was necessary to testify to the universality of the effect of the flight.

Scientific proceedings established flight as a phenomenon mastered through a precise chronology, an accurate vocabulary and testimonies of the exact time of the ascent and names of the travellers. This was the result of the visible gestures around the balloon, which staged various technical operations whereas reflecting a certain division of labour between the entrepreneurs, subscribers, and workers. The reports present a precisely controlled experiment, and express the magnificence of and astonishment at the skills involved. The reality was, however, full of small failures, setbacks and anecdotes. In Nantes, wind and the uncontrolled actions of people around the machine led to the destruction of nearly all the scientific measurement instruments that should have been brought on board.[64] Leveque would later carefully manage all the operations of the second experiment by himself to avoid such inconveniences. Most of the pictures tell another story of a balloon several inches above the stage, on the verge of being launched, the envelope completely filled by an "air lighter than air" and under the eyes of a huge, ecstatic crowd. They also depict the social materiality of the experiment which distinguished the groups near the balloon from the wider public: subscribers, those who had paid or were invited, sometimes divided into several circles; then, at some point beyond,

---

63   "... parlant, jabotant d'air, de gaz inflammable, citant sans cesse et prônant indistincte-
      ment les noms de Montgolfier, de Faujas, de d'Arlandes, de Pilastres, de Charles et Robert,"
      Mettra, *Correspondance secrète*, De Paris, Le 2 décembre 1783, 252- 257.

64   *Description faite de la seconde expérience aérostatique faite à Nantes*; Marie Thébaud-
      Sorger, "La mesure de l'envol à la fin du XVIIIe siècle. Les premiers ballons: affaires
      d'opinions ou d'exactitude?," *Histoire et Mesure,* 21 (June 2006): 35–78.

an undistinguished human gathering.[65] The sources are explicit: the majority of the city's inhabitants, from all levels of society, attended the flight. While this scenography tends to push a non-paying public to the outer the area of the stage, many opportunities existed to share the entire drama and catch the essence of the experiment. First of all, the raised stage allowed a better vantage point for the audience far away from the balloon. Second, many found an elevated viewpoint by perching on walls or roofs, while others used telescopes. Finally, the meaning of the experiment lay in the possibility of producing a universal consensus. Canon shots punctuated the different steps – when the balloon began to be filled, when it was full, and when the rope was about to be let out – in order to give a common understanding of the drama. In Nantes "the first vitriolic and zinc was poured in, announced by a canon shot" at 4:47am, but a leak appeared which needed to be repaired, and the process started again at 7am, but the balloon of 30 feet in diameter was only filled at 6pm. Thus, while the event offered many vantage points of view, easily locatable single elements were perceivable to all in attendance.

Audiences were reframed into a collective and unique creation: a public of citizens who consented to the new order of progress proposed by the experiment. While this reveals the different chapters of a central tale that everyone could tell, at the same time it produces various possible appropriations of the processes. The flight is a distraction. Above all, the depiction of the machine and its materiality made the phenomenon of the flight intelligible: the power of the flight lay in a volume of air displaced; the balloon rose to a certain elevation thanks to the lightness of the fluids it contained, and reached an equilibrium with the rarefied atmosphere; each component of the machine – soft fabric, basket, ropes, anchor (for landing) ballast, all apparatus taken on board, even the weight of the aeronauts – formed an addition which was subtracted from the global volume of air the envelope could contain. The difference defined the possibility of the ascent in an excess of lightness. For the *Suffren*, the calculation was the following: "weight of the aerostat + varnish + valve + net + the coach with seats + instruments and travellers' supplies = 12,500 cubic feet of inflammable gas = total of 955 pounds 12 ounces, upward lift = 20 pounds." Total weight for equilibrium = 975 pounds 12 ounces.[66] Daily newspapers and small leaflets were full of tables that allowed people to make additions and calculate this proportional relationship according to the size of the machine.

---

65   *Vue perspective de l'aérostat le Suffren par Hénon, représentant l'envol du ballon à hydrogène de Nantes, le 14 juin 1784*. Musée Dobrée, Nantes. (This engraving mentioning the Manufacture of varnish, and could serve as a sort of "trade card" for the entrepreneur).

66   *Description faite de la seconde expérience aérostatique faite à Nantes,* 12.

Balloon ascents were "*à la portée de tous le monde*" ("accessible to all"). For this purpose, and owing to the fact that ascents brought about a form of radical equality among those who remained on earth, all the different steps of the narrative should be perceived in terms of the whole assembly, even for the less educated crowds or those who are remained very far from the stage. By examining a few cases, especially in Nantes, we see that when a balloon took flight a chain of actions and operations unfolded in a specific order and under the watchful eye of notables and the remaining inhabitants of the city. Public evidence is part of the success of the experiment that otherwise would have held little meaning. And it implied that a soft, hermetic envelope filled with an air lighter than air produced by chemical reaction (combustion or gas production) gradually filled up, until the difference in weight altered the balance of the machine and it took flight and rose irrepressibly – the steps leading up to the flight were more or less controlled by the group surrounding the machine, watched by the crowd. The knowledge shaped through this staging was also shaped by individuals' own intellectual and social horizons. It was the result of human actions which were easy to understand, and therefore prodigious.

3　Conclusion

When Anna Cradock visited the cannon factory of Indret, next to Nantes, she took great pleasure in observing "the molten cannon taken to a water mill which drills and polishes it." This admiration was however tinged by sadness, for a technical achievement "dedicated to kill one's own kind." Far from an unquestioning consumer or ignorant female, she cast an expert eye on the production processes and artefacts but her understanding was not separated from a moral judgement on the social utility and ultimate use to which its products were put. Back in Nantes, she attended the next days at two lectures on balloon inflation where she shown great interest "regarding what concern the different qualities of airs" and enjoyed discussing at length with a lecturer she found "highly-educated" (probably the apothecary Louvrier). If there is no evidence that Ms. Cradock made a small balloon herself, the technological market surrounding ballooning presented a device that everyone could reproduce, at home or for friends. Pamphlets and "how-to" leaflets, small models, recipes for making varnish were widely available and "à la portée de tous le monde."[67]

---

[67]　Marie Thébaud-Sorger, *Une histoire des ballons. Invention, culture matérielle et imaginaire,1783–1909* (Momum: édition patrimoine), Collection « Temps et espaces des arts », Avril, 2010.

These brought people to cutting-edge knowledge on chemistry, gas, and heat at a crucial moment when the theoretical framework of phlogiston was being debated. Improvers of all kinds proposed reducing the cost of machines by substituting fabrics, or a new varnish, or another gas, or varying the method of gas production (distillation of vegetal matter and coal was attempted), as well proposing new methods of steering the machine. Observations made while attending a flight or taking part in experiments sparked more ideas, not only from educated or elite audiences, as each person was able "to translate in his or her own way what he or she perceives through a play of 'association and dissociation.'"[68] Linking the outcomes of science and technology to social and political speculation also helped to forge a public sphere in which everyone could legitimately claim to participate. However, institutions and scientists would ultimately take control of this realm, putting an end to a period of wider participation. The reconstruction of the material reception of early aerostatic flights, which has only been possible for historians because of the large number and variety of archives, visual and printed testimonies, newspapers, letters, manuscripts, and police reports, has informed us about the different layers that constituted the mediation process of such events, their different meanings, themselves shaped by their contextual reception. It also reveals what innovative French territory was capable of, especially in chemistry,[69] and how scientific and political involvement forged a peculiar public event. The enthusiasm for the aerostatic flights would take on a very different form in Britain or German lands.

At the end of the Enlightenment, technical experiments fostered the imagination and creativity of a whole community of improvers. Spectacles of technology became common process for entrepreneurs, craftsmen and inventors and formed the basis of a culture, which in turn transformed the wider urban culture. Such displays played a part in a growing interest in technological processes, and fostered the new materiality of the elements at stake, such as air, fire, and heat that could be mastered and reproduced artificially, thereby changing the relationship between societies and their natural environments. The "prodigies" of technology reframed public curiosity under new categories and expectations (in line with industrial changes). To build a methodology that allows us to scrutinize the space where these mediations were staged as a socio-material device (encompassing objects and people),[70] helps us to qualify

---

68   Rancière, *Le spectateur émancipé*, 23

69   John Perkins, "Chemistry Courses, the Parisian Chemical World and the Chemical Revolution, 1770–1790," *Ambix* 57 (2010): 27–47.

70   Noorte and Lezaun, "Materials and Devices of the Public."

better the infinite variations between curiosity and useful knowledge. As we have seen, this involved numerous and varied forms of negotiations, adjustments and accommodations. Each of these could varied from state to state. This kind of historical approach also helps us, by challenging the distinction between passive and active audiences, to resituate the "emancipated spectator" in the long genealogy of physical encounters between science and public, from popular science of the nineteenth century, popularization that aimed to instruct large audience to the new issues of "citizen science."

CHAPTER 8

The Space Between: James Dinwiddie and the Transit of Science, 1760–1815

Larry Stewart

By the end of September, 1793, the sometime Scottish schoolmaster and itinerant lecturer James Dinwiddie found himself standing in the court of the Chinese Emperor. This was as different a scene from his origin nearby Dumfries' docks, on the river Nith, as one may now imagine. Leaving Portsmouth for Pekin in 1792, his sea track had taken most of a year, amid the supercargo of Lord George Macartney's embassy sent to convince the Chinese of the benefits of permitting British trade beyond the limits of the existing foreigners' factory at Canton. Measuring by sextant and by the new chronometer the entire way, Dinwiddie was neither diplomat nor merchant, enlisted simply to help the British win more of the earth. This was a much-debated moment which, depending on whom you read, reflected badly on Macartney's failure or, otherwise, on the studied indifference of the Chinese who, according to some accounts, "never valued ingenious articles, nor...the slightest need of [British] manufactures" however brilliant.[1]

But here was a confusion of objectives. Whether ingenious, entertaining, or even useful, knowledge travels in many forms. We have long been accustomed to books and pamphlets, to letters and lectures and, these days, various forms of incessant media, most notably now Facebook, Twitter and their infernal company. This all now assumes traffic in popular accounts of ideas

---

1 See Linda Colley, "Britishness and Otherness: An Argument," *Journal of British Studies* 31 (October, 1992): 309–329, at 310. Cf. Earl H. Pritchard, *Anglo-Chinese Relations During the Seventeenth and Eighteenth Centuries* (1929; New York: Octagon Books, 1970), esp. chapter X; Alain Peyrefitte, *The Immobile Empire* (New York: Vintage, 1992); Benjamin A. Elman, *On Their Own Terms. Science in China, 1550–1900* (Cambridge, Mass., and London: Harvard University Press, 2005), at xxxiv-xxxv; Elman, "Who is Responsible for the Limits of Jesuit Scientific and Technical Transmission from Europe to China in the Eighteenth Century?", in Clara Wing-Chung Ho, ed., *Windows on the Chinese World. Reflections by Five Historians* (Lanham, Boulder and New York: 2009), 59; and Peter J. Kitson, *Forging Romantic China. Sino-British Cultural Exchange 1760–1840* (Cambridge and New York: Cambridge University Press, 2013). On the traces of Macartney, see Robert Swanson, "On the (Paper) Trail of Lord Macartney," *East Asian History* 40 (August, 2016): 19–25.

and images whatever the limit of their substance. I propose a slightly differ-ent perspective on the trade in ideas. I refer not just to commerce in material goods, but to those reflecting scientific notions which defined the great reach of European Enlightenment. Such circulation, it may be argued, "opens the way for new kinds of historical actors" beyond royal societies or select agents of the Crown.[2] In the late 18th century they were many especially in the con-federacy of experimental practitioners across cities, regions and nations, par-ticularly in the "global networks of power, commerce, and knowledge."[3] Some may now recognize lecturers and purveyors of scientific goods like Benjamin Martin, William Jones, John Canton, James Short, John Aiken, James Ferguson, Adam Walker, and *many* others revealed in a few scattered traces in archives and attics. Instruments and their agents carried information and, by inducing translation across spaces, scientific devices transformed localities in unique encounters of natural knowledge and conquest of imperial spaces. What fol-lows is but one curious example, evidence of which has miraculously survived in long-hidden hints of the scientific life of the largely unknown Dinwiddie, once a maths teacher in the small market town of Dumfries, from mid-18th century Scotland, but who manufactured an everyday life out of experimental instruments. His origins were relatively modest, born in 1746 but, like many Scots, he was carried by the tides of the 18th-century British Empire in ways he surely never intended. His was one of those many stories, many lost, overshad-owed by those greats who filled the shelves with books and ideas. Dinwiddie's personal and mostly marginal trajectory can tell us much about how the early-modern scientific world functioned.

The transport of instruments and scientific goods across vast oceans tra-versed a critical intellectual space between origin and reception. What mean-ing could experimental devices have to those in foreign places which had no sense, beyond rumour, of an Isaac Newton or a Joseph Priestley? If a mech-anism or piece of apparatus meant the same thing in Canton as it had in London, then perhaps that might demonstrate an imperial, natural, or even universal scientific lexicon which in turn reflected an imperial claim to the command of nature. But Chinese reactions to supposedly brilliant European

---

2   Pedro M.P Raposo, Ana Simoes, Manolis Patiniotis and Jose. R. Bertomeu-Sanchez, "Mov-ing Localities and Creative Circulation: Travels as Knowledge Production in 18th-Century Europe," *Centaurus* 56 (2014): 167–188, at 168.

3   See Fokko Jan Dijksterhuis and Andreas Weber, "The Netherlands as a Laboratory of Know-ing," introduction to Dijksterhuis, Weber and Huib J. Zuidervaart, *Locations of Knowledge in Dutch Contexts* (Leiden: Brill, 2019): 1–14; and Simon Naylor, "Introduction: historical geogra-phies of science—places, contexts, cartographers," *British Journal for the History of Science* 38 (March, 2005): 1–12.

FIGURE 8.1   Hahn/Vuilliamy planetarium, by William Alexander, 1793. Permission of the
British library.

instruments, like the *Weltmaschine* or planetarium of the German Pietist
Philipp Hahn, reworked in London by the British maker Justin Vulliamy and
taken by Macartney to Pekin, could hardly suggest all mechanisms were
viewed everywhere with consistent expectations (figure 1). The wonder of the
Weltmaschine, with its gilt and glitter, was dismissed in Pekin as a trifle.[4]

Apparently it was once thought so remarkable that the repairs to the plan-
etarium made in London by Vuillamy were enough justification of the skill of
British instrument makers to impress the potentates of Britain's Board of Trade.
But there is another level at which an apparatus designed to keep celestial time

---

4   Kitson, *Forging*, 148–149; See Simon Schaffer, "Instruments As Cargo in the China Trade," *History of Science* 44 (2006): 217–246, esp. 221, 227, 232; Simon Schaffer, "Easily Cracked. Scientific Instruments in States of Disrepair," *Isis* 102 (2011): 706–717, at 715–716

FIGURE 8.2   Hindostan, by Thomas Luny, c. 1793. Permission of the National Maritime
Museum, Greenwich.

might also be construed. This was the age of the instrumental solution to the
conundrum of longitude at sea by John Harrison's marine chronometer. We
know that Macartney's expedition had at least two chronometers on board,
one employed by James Dinwiddie throughout the voyage.[5] What, after all,
could be more impressive than such long voyages measured more precisely
than ever possible, simply by combining calculations of time and distance in
one device? How could this induce a shrug?

There was surely nothing simple about the planetarium, and certainly not
of the much smaller and sophisticated marine chronometers. Could even
more narrowly refined apparatus used in the practical measurement of things,
like the timing of the tides or eclipses, or even in revealing the chemical reactions of dyes in Wedgwood mortars, assert the mastery of universal, natural
knowledge? Instrumental cargoes like this were thus not only goods in transit,
certainly not those in the holds of HMS *Lion*, the East India Company ship *Hindostan*, or the brig *Jackall* driven east by the "Trades" (figure 8.2).

5   See here, *Fisher's Colonial Magazine* 4 (September-December, 1843): 404–414; from British
    Library, Ms. IOR G/12/20, ff. 596–618; Maxine Berg," Macartney's Things. Were They Use-
    ful? Knowledge and the Trade to China in the Eighteenth Century", https://www.lse.ac.uk/
    Economic-History/Assets/Documents/Research/GEHN/GEHNConferences/conf4/Conf4-
    MBerg.pdf; Berg, "Britain, industry and perceptions of China: Matthew Boulton, 'Useful
    Knowledge' and the Macartney Expedition to China 1792–94," *Journal of Global History* I
    (2006): 269–288.

Machines and devices might be carried, whether elaborate orreries, or pottery from Josiah Wedgwood together with his beautifully bound catalogue, suggesting Midlands' porcelain might be sent to China, along with his clay pyrometer much in demand in Britain for measuring temperatures in the kiln. Likewise, amidst the vast manifest of machines and models there were those of Birmingham's Matthew Boulton, experimental apparatus from London instrument makers like Thomas Blunt and Edward Nairne, a set of pulleys from the engineer John Smeaton, achromatic telescopes from John and Peter Dollond, Jesse Ramsden's measuring instruments and theodolites, along with chronometers and electrical machines.[6] This partial list is worth noting as the same objective had been explicitly intended even in an earlier, and aborted, expedition in 1787 under Charles Cathcart who died on route to China. Much of that apparatus was returned to Britain with the fleet and lay in storage until consigned to Macartney when the Board of Trade ordered them repaired and restored in 1792.[7] What began with great confidence, those objects transported ideas and images of British superiority as though machines were a measure of a civilization resulting, perhaps inevitably, in confusion and contested reputation.[8]

James Dinwiddie was then along to impress—to employ an array of apparatus refracting the best of British achievement.[9] Following such a long voyage, after leaving Macao, skirting the foreign factories at Canton and coming ashore off the Yellow Sea, Dinwiddie was soon dismayed. Up river, he wrote, "Nothing observable of much scientific knowledge transpired."[10] Yet he still held out hope for their final destination of Pekin. Dinwiddie was ultimately appalled at the lack of interest shown in his brilliant machines, some having been the focus of his decades as itinerant lecturer, but now ignored by mandarins studiously intractable.[11] This clearly surprised him. Could this be reduced, as is

---

6   British Library, India Office Records, IOR G/12/20, fols. 595–618. Articles bought for Macartney expedition January-August, 1792. It was through Smeaton, among others, that Dinwiddie gained access to meetings of the Royal Society in 1790. Royal Society, JBO/34 Journal Book (1789–1792), May 13, 1790. I owe this reference to Kristen Schranz.

7   British Library, India Office Records, IOR G/12/20, fols. 638–641.

8   Cf. Lewis Mumford, *Technics and Civilization* (1934, New York: Harcourt Brace & World, 1962); Michael Adas, *Machines as the Measure of Men: Science, Technology, and Ideologies of Western Dominance* (Ithaca: Cornell University Press, 2015).

9   Peyrefitte, *Immobile Empire*, 138–139.

10   William Jardine Proudfoot, *Biographical Memoir of James Dinwiddie. Astronomer in the British Embassy to China, 1792, '3, '4, Afterwards Professor of Natural Philosophy in the College of Fort William, Bengal. Compiled from his Notes and Correspondence by his Grandson,* William Jardine Proudfoot (Liverpool: Edward Howell, 1868), 41.

11   Peyrefitte, *Immobile Empire*, 276, 287.

often suggested, to an inevitable clash of cultures or conquest gone awry? The Chinese had good reasons to keep Europeans confined to Canton. No amount of impression was likely to give the British an exclusion from long-standing limits. Devices would not be traded for the keys to escape Canton. Dinwiddie had long struggled to make his way by entertaining the curious public with the latest science throughout Scotland, Ireland and in England, from Midlands to spa towns, to London's metropolis where, at least since the end of the 17th century, large audiences had often subscribed great sums to be entertained by experimental conjurors. By comparison, Dinwiddie found the lack of curiosity in Pekin inexplicable.

Pekin surely was an entirely alien world. At the Emperor's Court, objects conquered no one. When shown apparatus Dinwiddie deemed important, on October 4, the Emperor supposedly sneered, "These things are good enough to amuse children." Of the larger court, among the mandarins and eunuchs, there was no acknowledgement when Dinwiddie demonstrated the power of lenses to ignite wood and melt metal, even in a cylinder devoid of air. Such indifference, he wrote, "only proved that a Chinaman will entertain ideas truly provoking a European philosopher."[12] The next day Dinwiddie recorded that "few experiments, comparatively speaking, had been wrought, and these were entirely lost on the most prejudiced of people."[13] Most offensive to him, above all, was the sudden dismantling the very next day and behind his back of the housing of a great burning lens made by William Parker of Fleet Street. Dinwiddie was shocked: "That lens—of which there is not an equal in the world—is consigned to everlasting oblivion."[14] Such attitudes established the useful legend of obstruction the British erected around the apparent failure of their embassy. Yet, all was not quite so straightforward. Dinwiddie was not so obtuse to assert all could be reduced to Chinese obstinacy. He admitted the British were hardly best prepared to secure the desired impression. He complained, "What information could we derive respecting the arts and sciences

---

12 Proudfoot, *Biographical Memoir*, 53; Bernard Lightman, Gordon McOuat and Larry Stewart, eds., *The Circulation of Knowledge Between Britain, India and China* (Leiden and Boston: Brill, 2013), 5–6.

13 Quoted in Peyrefitte, Immobile Empire, 297; See also Benjamin Elman, "China and the World History of Science 1450—1770," *Education about Asia* 12 (2007): 40–44, on p. 43.

14 Proudfoot, *Biographical Memoir*, 54; Peyrefitte, *Immobile Empire*; Cf. William Jardine Proudfoot, *'Barrow's Travels in China.' An Investigation into the Origin and Authenticity of the 'Facts and Observations' Related in a Work Entitled "Travels in China, by John Barrow, F.R.S."* (London: George Philip, 1861), 34–40. On Parker's lenses see W.A. Smeaton, "Some large burning lenses and their use by eighteenth-century French and British chemists," *Annals of Science* 44 (May, 1987): 265–276.

in a country where we could not converse with the inhabitants?"[15] In fact, it was then neither new knowledge nor praise the British sought, but access. That tarnished all.[16] This was not simply confusion over commodities, as of a cargo cult. Even with the best intentions, knowledge did not always travel well. Rather, Dinwiddie's dismay was a reflection of the impermeability of ideas, of the difficulty of translation of the learning inherent in the goods themselves. Machines might carry knowledge, but ideas needed a translator, or demonstrator like Dinwiddie, in *the new spaces* it had taken months to reach.

1   A Matter of Instruments

This episode, especially the response at the Emperor's court, raises a significant question beyond the ease of explanation that suited a British legend. After all, Dinwiddie's role may not have settled on him entirely comfortably. He had a long history with devices of impression. Yet, he also knew that instruments were more than machines of amusement. I propose here a picture of intellectual goods that will take us beyond a common utility or claim to master nature especially when, in the late 18th century, the role and importance of experimental, natural philosophy and improvement was actually much contested. Instruments were then the key to many a career. A presumption of useful knowledge had once been adopted by the mathematician of Dumfriesshire. Through a Baconian lens, the thriving port of Dumfries could have much more to offer than the apparent vice and temptations of much larger cosmopolitan and academic cities like Edinburgh. And James Dinwiddie would thereby, in ways unimaginable to him, establish a reputation much farther afield. But, thirty years before China, he could not have known the tide upon which he would set when he once made a proposal for a subscription in July, 1765, to gentlemen of the town in an effort to secure for the school an orrery.[17] He was

---

15   J.L. Cranmer-Byng, ed., *An Embassy to China. Being a journal kept by Lord Macartney during his embassy to the Emperor Ch'ien-lung 1793–1794* (Hamden: Archon Books, 1963), 53, 266, 347; Lightman, McOuat, Stewart, eds., *Circulation of Knowledge*, 4–5.

16   Cf. Maxine Berg, "Macartney's Things"; Berg, "Britain, industry and perceptions of China"; Berg, "The Asian Century: The Making of the Eighteenth-Century Consumer Revolution. Cultures of Porcelain between China and Europe," www.lse.ac.uk/economicHistory/Research/GEHN/GEHNPDF/Confio_Berg.pdf.

17   See L. Stewart, "The Spectacle of Experiment. Instruments of Circulation, from Dumfries to Calcutta and Back" in Bernard Lightman, Gordon McOuat, and Larry Stewart, eds., *The Circulation of Knowledge Between Britain, India and China. The Early-Modern World to the*

not about to stop there. In 1766, he further recommended the governors of the school purchase a "proper Philosophical apparatus" for a course of natural and experimental practise useful, perhaps, for boys launching into trade and navigation. This was not, in the growing realm of philosophical lecturers, entirely a novelty. Others were attempting the same thing as, for example, the teacher and publisher Benjamin Martin who ended his days as an instrument-maker, bankrupt from competition, and suicidal in London (figure 8.3). Dinwiddie, who evidently knew Martin, asserted the principle that "... Science is not only allowed to be ornamental to the Gentleman & Scholar, but known to be strictly connected with the most common occupations & most useful arts." For the consideration of the town worthies, he once argued, in a Baconian tradition, that experimental science "assists the industry of the Merchant, Farmer & Mechanick."[18] This was a refrain copied well into the next, industrial, century.

Observational instruments and mathematics might seem the servants of any trading town. Indeed, for the next decade, Dinwiddie promoted the virtues of devices and experiments alongside mathematics for boys. In 1774, we would have found him in London selecting, presumably off the shelf, devices from 'Mr. Martin' while he set about buying the works of Priestley on airs and of the lecturer James Ferguson on astronomy.[19] The connection to Martin is interesting on several levels, not the least being Martin's own origins as an instructor in natural philosophy, his trajectory into the alleys of London's instrument makers along the Strand and Fleet Street, and even his failure to secure anything but studied hostility from the guardians of a gentlemanly science in the Royal Society.

As Simon Schaffer once remarked, "The market threatened moral order but helped controversies close by generating a standard range of instruments, experiments and language with which to describe them."[20] Apparatus was crucial to the public philosopher, to seduce an audience, especially for those performers like Dinwiddie who deployed an array of instruments and which

---

Twentieth Century (Leiden and Boston: Brill, 2013), 21–44, esp. 24; Dalhousie University Archives (DUA), MS 2–726, A 32, List of the Gentlemen Subscribers for purchasing a Philosophical Apparatus, for the Use of the School of Dumfries...[poss. 1774]. A copy of the early subscription for the orrery is in the private collection of Chris Green, dated July 16, 1765.

18   DUA, Dinwiddie fonds, Dinwiddie, "Plan of Education," dated September, 1766.

19   DUA, MS 2–726, F 1. Early Experiments, ca. 1774 (September, 29).

20   See Simon Schaffer, "The Consuming Flame: Electrical Showmen and Tory Mystics in the World of Goods," in John Brewer and Roy Porter, eds., *Consumption and the World of Goods* (London and New York: Routledge, 1993), 489–526, esp. 512.

FIGURE 8.3
Benjamin Martin, engraving
by R. Page, 1815. Permission
of the Science Museum

induced, for him, both a dependency and an obsession. But the influence was
not Martin alone. Even while crafting his own devices, Priestley in the Midlands was also obtaining apparatus from others. It was laboratory practice that
the chemist Priestley chose to promote. He argued, "the expense of instruments may well be supplied, by a proportional deduction from the purchase
of books, which are generally read and laid aside, without yielding half the
entertainment."[21] Dinwiddie soon accepted Priestley's own blandishments of
the ways to experimental knowledge:

---

21   Joseph Priestley, *The History and Present State of Electricity, with Original Experiments*.
      Second edition, corrected and enlarged. (London, 1769), x. On the expense of apparatus
      see Maurice Crosland, "A Practical Perspective on Joseph Priestley as a Pneumatic Chem-
      ist," *British Journal for the History of Science* 16 (November,1983): 223–238, esp. 233–234

It were much to be wished, that philosophers would attend more than they do to the construction of their own machines. We might then expect to see some real and capital improvements in them; where as little can be expected from mere mathematical instrument makers; who are seldom men of any science, and whose sole aim is to make their goods elegant and portable.[22]

Dinwiddie took this to heart. Yet, without private wealth, instrumental investment was as much affliction as promise. For novices in philosophical lectures, experimental demonstration was clearly refracted, as Jim Bennet has shown us, through devices increasingly available from the shop shelf.[23] But, as Dinwiddie often worried, his finances imposed severe limits. His growing attraction to experimental devices led to a career that was endlessly itinerant and very much conflicted over the connection between use and enlightenment. Dependent on attracting local patronage, he also mused, in 1767, that "the dignity of human nature eminently point out something further than what is mechanical and seems to be adapted for the entertainment of more refined knowledge than the mere concerns of the groveling business of life."[24] In a gentlemanly culture, utility might appear to contest the cultivation of enlightenment. This was one of the essential tensions of 18th century Britain—where empire and experiment did not always easily converge even on a Baconian tide.

Dinwiddie was driven by the growing fascination with endless invention and endless experiment, renown in the British enlightenment. In desperation induced by his instrumental addiction, he was forced to look beyond the local school or even his connections in Edinburgh University from which he obtained an M.A in 1778. At a crossroads, his sought out a fellow philosopher in the chemist Thomas Percival of Manchester who referred him to William Cullen in Glasgow.[25] But there proved few ready avenues of patronage or

---

22   Joseph Priestley, *The history and present state of electricity, with original experiments* (London: printed for C. Bathurst, and T. Lowndes, in Fleet-Street; J. Rivington, and J. Johnson, in St. Paul's Church-Yard; S. Crowder, G. Robinson, and R. Baldwin, in Pater-Noster Row; T. Becket, and T. Cadell, [1775]), 83. For an exhaustive treatment of creating apparatus see Simon Werrett, *Thrifty Science. Making the Most of Materials in the History of Experiment* (Chicago and London: University of Chicago Press, 2019).

23   Jim Bennett, "Shopping for Instruments in Paris and London," in Pamela H. Smith and Paula Findlen, eds., *Merchants & Marvels. Commerce, Science and Art in Early Modern Europe* (New York and London: Routledge, 2002), 370–395.

24   DUA, MS 2–726, K 1. Statement Concerning a Course of Natural Philosophy (taught by Dinwiddie), dated March, 1767.

25   Glasgow University Library, MS Cullen 99 (1). Thomas Percival to Cullen, November 11, 1776?

promotion. Dinwiddie left Dumfries to venture possibilities as itinerant philosopher, recruiting chances in a culture overwhelmingly dependent on connection—before it was itself overwhelmed by industrial speculation. His track took him to Leith and Dundee, and then to Ireland where, in a span of very few years he was delivering lectures in Lisburn, Newry, Drogheda, Waterford, Dublin and Cork.[26] Here he first encountered, likely among his auditors, Richard, Viscount Wellesley who would go on to become Governor-General of India in 1798 and founder of Fort William College in Calcutta.[27] As we will see, this was one of the few connections which would serve Dinwiddie well. While in Ireland, he became increasingly aware of competing lecturers having received from Rob Stevenson, a bookseller in Newry, a copy of the syllabus of Adam Walker, itinerant lecturer originally of Manchester, who may have provided something of a model.[28] It was not long before Dinwiddie traded the Celtic connection for larger audiences in England. By December 1783, he was involved in the launch of a hot air balloon in London at Buckingham Gate, followed by the same spectacle in Bristol in January, and Salisbury in March. He was soon back in Dublin with the same display.[29] However, this was simply a stop on his many travels while his associations in England began to bear some, if still uncertain, fruit. He established links to the manufacturers of the Midlands and these, in turn, would give him access to the corridors around Joseph Banks, the Royal Society and the Board of Trade in London. This was his critical introduction to the Banksian Empire.

Dinwiddie's contacts were hard won. It seems that by 1790 he was contemplating lecturing at Richmond, nearby the Royal Court. Here he got assistance from Mrs. Dundas (likely the wife of Henry Dundas of the Board of Trade), along with that from a Mr. Ewart who once had a connection to an earlier failed effort to cultivate the Chinese. This link was likely Dr. John Ewart, an early proponent of pneumatic medicine. He was the brother of Peter, the Birmingham engineer, educated at the Dumfries Free School who was to be employed by Matthew Boulton from 1788, and by the Boulton-Watt firm at Soho in Birmingham from 1790.[30] While the initial entrée to the English political establishment

---

26   Proudfoot, *Biographical Memoir*, 4.

27   Stewart, "The Spectacle of Experiment," 39; and Jan Golinski, "From Calcutta to London: James Dinwiddie's Galvanic Circuits," in Lightman, McOuat, and Stewart, eds., *The Circulation of Knowledge Between Britain, India and China*, 75–94, esp. 79.

28   DUA, MS 2–726, A 92, Rob Stevenson to Dinwiddie, October 21, 1782.

29   *Morning Post*, December, 1783; *St. James's Chronicle*, January 31, 1784 and March 13–16, 1784; Stewart, "Spectacle of Experiment," 35.

30   DUA, MS 2–726, A 33. Mrs. Dundas to Dinwiddie, July 1790, on a list of subscribers for his lectures at Richmond. On Peter Ewart see *Memoirs of the Literary and Philosophical Society of Manchester*. Series 2, vol. 7 (London: John Weale, 1846): 114ff.

was as incidental as it was uncertain, we know that Dinwiddie was lecturing around London from at least December, 1789.[31] Joseph Banks, P.R.S., and Lord Macartney both consulted him by the spring of 1792 on the Embassy then being organized by Dundas to the court of the Chinese Emperor where, as we know, the cultural limits proved most complex and for which the British proved unprepared.[32]

Despite all appearance, China was not entirely the disaster we might have assumed from the common accounts. Dinwiddie's imperial disappointments were transformed by unexpected opportunity. Dinwiddie was sent off with some of the valuable instruments refused by the emperor's court, given to the Scottish philosopher by Lord Macartney "as part payment for his service..."[33] Setting out on the long journey for London, he was first directed to Calcutta to deliver tea plants for the botanic garden.[34] This layover ultimately lasted another 12 years until he finally cut himself free of the East India Company and a temporary teaching post in the College of Fort William. But, as Charles Withers has pointed out, for enlightenment travellers like Dinwiddie, especially for Scots who populated the empire, "geographical mobility was also social mobility."[35] In crossing geographic and social spaces we can follow the transit even of imperfect knowledge. In samples and in instruments, scientific ideas travelled to and from India, between the shops around London's Fleet Street in particular and Calcutta's Fort William. Once again Dinwiddie turned to the lecturing trade. Shortly after his arrival in Calcutta, with Macartney's apparatus and some of his own, he set up as a public demonstrator even as he had earlier made uncertain way. He had some reputation well before 1792 having given numerous lectures on natural and experimental philosophy, notably on

---

31   DUA, MS 2–726. A 91. William Steven to Dinwiddie, December 16, 1789; See also BL Burney Collection. *The World*, November 20, 1790; *Morning Chronicle*, April 1, 1791.

32   Cornell Library, Wason Collection, volume 4. Dinwiddie to Macartney, date unclear, Saturday 10 June [1792?]; See also Macartney to Banks, April 20 & April 26, 1792. *The Indian and Pacific Correspondence of Sir Joseph Banks, 1768–1820*, ed. Neil Chambers (London: Pickering & Chatto, 2007) III, 370–371. My thanks to Jordan Goodman for these references. Cf. Maxine Berg, "Macartney's Things," 10ff.; Berg, "Britain, industry and perceptions of China," 269–288, esp. 274.

33   British Library, India Office Records, 354.54. R. H. Phillimore, *Historical Records of the Survey of India*, Volume II (1800 to 1815). (Dehra Dun, India: Offices of the Survey of India, 1950), 251, 395.

34   Peyrefitte, *Immobile Empire*, 393–395; Kitson, *Forging*, 146–147; Minakshi Menon, "Medicine, Money, and the Making of the East India Company State: William Roxborough in Madras, c. 1790," in Anna Winterbottom and Facil Tesfaye, eds., *Histories of Medicine and Healing in the Indian Ocean World*, I (Basingstoke and New York: Palgrave MacMillan, 2016), 151–178.

35   Charles Withers, *Placing the Enlightenment. Thinking Geographically about the Age of Reason* (Chicago and London: University of Chicago Press, 2007), 229.

chemistry and electricity then very much in vogue in Europe. And, by 1795, the Board of Trade was ordered to consult the philosopher in Calcutta "on several points of Chemistry Mechanics and Natural Philosophy which have relation to the business of their department."[36] Through the East India Company, London's long reach found him there.

At the end of the 18th century new truths emerged, out of experiments on airs and laboratory glassware as much as from new worlds encountered throughout the vast web of empire. Dinwiddie became a purveyor of devices, many he had bought and much self-constructed, a peripatetic and chance philosopher amidst the breadth of imperial agencies.[37] From modest origins, in an age of connection, he depended on links within the eighteenth-century state and among merchants who exploited it. Most useful were agencies of the Board of Trade directed by Henry Dundas, and the East India Company links that Banks happily manipulated through the threads strung from Kew.[38] China, hence, was a brief sojourn until Dinwiddie's crossing over the docks of Calcutta, at the end of September, 1794, allowed him to exploit talents honed since he left Dumfries in the 1770s. Long before he had embarked from Portsmouth he was, as we now know, successful enough to be brought to the attention of Banks and Macartney, launching him across oceans with his many notes in a sea chest.[39] These traces show, above all, that Calcutta made the difference. It was there, by the late eighteenth century, where science and apparatus were enlisted by the hierarchies of empire and industry to transform the world and to assert their European importance.[40] In India, with new objects to display, he found a certain audience and more reliable patronage than had long eluded him but which in the end would make his private fortune. This was so not merely because he was an entertainer, but because he could command experimental expertise, in an otherwise tumultuous age hungry for the

---

36   Dalhousie University Archives [DUA], MS 2–726. Dinwiddie Journals, Calcutta, B 21, July 13, 1795; July 23, 1795; University of Guelph Special Collections, Dinwiddie Correspondence, XS1 MS A164, no. 47. W. Emmerston? to Dinwiddie, January 31, 1797.

37   Cf. John Brewer, *The Sinews of Power: War, Money and the English State, 1688–1783* (Cambridge, Mass.: Harvard University Press, 1990).

38   Here I rely on the assistance of Jordan Goodman who provided me with a draft of his "Spot the Go-Between(s): Joseph Banks, Knowledge, and Interpreters of the Macartney Embassy to China, 1792." See also Kitson, *Forging*, 135.

39   See here Karen Smith, "The Dalhousie University James Dinwiddie Collection," in Bernard Lightman, Gordon McOuat, and Larry Stewart, eds., *The Circulation of Knowledge between Britain, India and China* (Leiden and Boston: Brill, 2013), ix–xii.

40   See Lissa Roberts, "Producing (in) Europe and Asia, 1750–1850," *Isis* 106 (December, 2015): 857–865.

FIGURE 8.4   Calcutta from Fort William, in William Hodges, *A view of Calcutta*, 1793.
Permission of the British library.

certainty natural laws could provide. Here he could translate knowledge across spaces from metropolis to frontier and, ultimately, back to London.

The most successful moment in Dinwiddie's remarkable trajectory was his stay in Calcutta (1794–1806) which delayed his return to Britain beyond expectation (figure 8.4). In India, he was able to secure a previous Irish connection with Lord Wellesley through the College of Fort William.

Both Wellesley and Sir John Shore, his predecessor as Governor-General in India, along with many well-connected East India Company traders, subscribed to Dinwiddie's experimental lectures over more than a decade in Calcutta. And serving the Board of Trade in Calcutta on the possibility of chemical sources for the indigo trade, Dinwiddie was asked to search out Indian manufactures. Thus, he communicated with Dr. Helenus Scott in Bombay on the production of vitriolic acid as well as on the uses of nitre in the treatment of disease.[41] His early Baconian notions were no passing enthusiasm. And Dinwiddie's ties, to the orbits of Banks and Macartney, of Dundas and the imperium, represented none of the aversion to the political establishment otherwise often challenged by republican and scientific contemporaries like Priestley or Ben Franklin.

---

41   University of Guelph, Special Collections, XS1 MS A164, no. 94. Helenus Scott to James Dinwiddie, August 23, 1796.

Through instruments and chemical goods Dinwiddie, even at such a distance, remained as close to London's science as he ever dared hope. The traffic
in ideas might modify the uncertainties and disappointments that frequently
afflicted the itinerant. Thus, experimental devices were a lifeline. He never lost
touch with London about books and apparatus, even when news was otherwise sporadic, untrustworthy and, in an age of revolution, frequently alarming.
There was a great difference between actively probing the laws of nature while
waiting upon ships arriving with uncertain gossip from Europe. Certainly, East
Indiamen afforded the means to get the latest apparatus, to enable the measure of natural things like electric shocks or distances to stars, and books to
show him how. As we shall see, immediately upon arrival in Calcutta, he was
in touch with suppliers in London. One of the most significant was surely the
chemist Joseph Hume of Long Acre, amid the denizen of instrument makers'
shops and scientific promoters nearby Covent Garden. Likewise, by November,
1795, Dudley Adams, son of the recently-deceased instrument maker George
Adams of Fleet Street, offered Dinwiddie in India several optical instruments
including a camera obscura ideal for public showmen. Two of these had
already been aboard Macartney's ships, probably rescued from the aborted
Cathcart expedition of 1787.[42] By the spring of 1796, following orders Dinwiddie had already placed, George Adams' widow herself fulfilled Dinwiddie's
request for books from remaining stock, employing the assistance of Hume
the chemical manufacturer. Trade connections proved critical as Hume would
regularly make use of the firm David Scott & Co, of both London and Calcutta,
to handle shipping on East India Company vessels.[43] This must have been a
regular occurrence, as surviving notes on sea voyages establish that Dinwiddie
kept studiously up to date with the latest of European scientific debates. Scientific information moved in imperial channels. By May of 1795, he had received
notice from Hume in London that answers as

> you may require will be found in the books which were ordered. I [Hume]
> find it impossible to get any information from the practical people here

---

42   British Library, India Office Records, G/12/20. Articles bought for Macartney expedition
     January-August, 1792, fol. 617v, fol. 640v.
43   DUA, MS 2–726. Dinwiddie Correspondence, A-1. Dudley Adams to Dinwiddie, November
     27, 1795; A-59. Joseph Hume to Dinwiddie, May 17, 1796. On Scott & Co see C.H. Phillips,
     ed., *The Correspondence of David Scott Director and Chairman of the East India Company
     Relating to Indian Affairs 1787–1805,* 2 vols. (London: Royal Historical Society, 1951); and
     Savithri Preetha Nair, "Bungalee House Set on Fire by Galvanism': Natural and Experi-
     mental Philosophy as Public Science in a Colonial Metropolis (1794–1806)," in Lightman,
     McOuat, Stewart, eds., *Circulation of Knowledge,* 60. Scott was a friend to Henry Dundas.

except from Horton, who wrote you every thing he knew in Sept. last. Horton also sent two books and two models of furnaces, which are in the box with the chemicals.[44]

This route continued in use for a decade, a track traced on the sea, ships' holds full of books and bottles filled by the chemical trade, secured in bundles with the *Encyclopaedia Britannica* (the 1797 Third edition with its expansive consideration of chemistry and electricity, soon difficult to obtain) and William Nicholson's *Journal of natural philosophy, chemistry and the arts* (published from 1797) (figure 8.5).[45]

When new knowledge is in vogue, like the chemical revolution, its grasp expands beyond the local and enthusiasm reforms the everyday.[46] This we now take as commonplace, as though the internet suddenly discovered global connection. But it was true also in the early-modern world, even when news travelled less quickly. Scientific objects as much as the written text effectively magnified a public philosophical interest and the Western presumption of the command of nature. One common instance at the farthest reach of empire was the fascination with making electricity. The phenomenon had, of course, attracted much attention throughout much of the prior century, both in Ben Franklin's mortal danger beneath a kite and, more helpfully, as medical devices designed by the likes of George Adams and Edward Nairne in London. By December, 1801, in Calcutta, Dinwiddie had already constructed his first Voltaic pile and, following the intelligence from Europe, was soon engaged in trials of the decomposition of water into its constituents of oxygen and hydrogen. It is striking he noted in his journal that, "in short the phenomena were exactly the same as described by William Nicholson in his first experiment on water."[47] But such phenomena were not new to Nicholson, even if his design was. By April, 1802, Dinwiddie constructed a battery of 120 pairs of plates, essentially based on Volta's discovery of the bi-metallic effect. In May, after having read reports obtained from London, Dinwiddie concluded that in

---

44 DUA, MS 2–726. Dinwiddie Correspondence, A-59. Joseph Hume to Dinwiddie, May 26, 1796. Possibly William Horton, frame maker, St. Matthew's, Bethnal Green. *Mechanic's Magazine* 47 (Jan.-June, 1847): 164.

45 DUA, MS 2–726. Dinwiddie Correspondence, A-59. Hume to Dinwiddie, August 29, 1800; Hume to Dinwiddie, July 8, 1802.

46 Cf. Simon Schaffer, "Easily Cracked. Scientific Instruments in States of Disrepair," *Isis* 102 (December, 2011): 706–717; Alexi Baker, "'Precision,' 'Perfection,' and the Reality of British Scientific Instruments on the Move During the 18th Century," *Material Culture Review* 74/75 (Spring, 2012).

47 DUA, MS 2–726. D22. Journal of Galvanism, March 19, 1802.

FIGURE 8.5
William Nicholson, engraving
by T. Blood after S. Drummond,
c. 1812. Permission of the
Science Museum picture
library.

the right conditions his pile would produce electricity 200 times more rapidly, although of low intensity, than that produced by common static generators or Leyden jars long familiar to experimenters like Joseph Priestley.[48]

Dinwiddie was intent on new electrical experiments and he, as always, needed ever more apparatus. He was determined to test the physiological effects of Galvanism. In the spring and summer of 1803, he was making notes on attempts to treat deafness including, as it happened, his own by application of shocks. This was followed later in the summer by musings on the celebrated efforts of Giovanni Aldini, Galvani's nephew, who claimed success in "several cases of insanity," in apoplexy, "and other disorders of the head," indiscriminate though they were. He had heard about Aldini's experiments on a decapitated dog so that he noted, "A curiosity was expressed to have the experiment tried on a criminal newly executed..."[49] Dinwiddie did not choose to include that trial in his adventures at least. By June he had noted the medical experiments

48   DUA, MS 2–726, D22. Journal of Galvanism, April 14 and May 1, [no. 20], 1802.
49   DUA, MS 2–726, D22, Journal of Galvanism, June, 1803.

of Dr. Bischoff of Jena and Jacques Nauche, President of the Galvanic Society in Paris, who had tried shocks in cases of paralysis and blindness. Like chemistry, electricity had crossed over the boundary from laboratory to the doctor's surgery.

Dinwiddie's connections came in the cargo of East India ships. By 1803, he heard from Hume in London about the trials of war-torn Europe. Britain was then highly alarmed. Hume warned, should Dinwiddie consider a return from India, that

> ... the Corsican upstart, who now tyrannizes over the greatest part of Europe, may perhaps urge us to become a nation of soldiers and you will find me in my laboratory with a bayonet on my side decomposing and compounding the various articles of our science.[50]

The armed philosopher in a London laboratory was an image to behold. Along with available articles in the instrument-makers' shops, including glass tubes ordered for Dinwiddie's experimental programme, Hume advised him on the latest electrical apparatus:

> You are no doubt already acquainted, that the galvanic pile as invented by Mr. Cruickshanks of Woolwich, is now generally employed in preference to the vertical pile of Volta—the alternative metallic plates are fixed [crossed out, 'horizontally'] in troughs made of wood and lined with a cement to render them impermeable to fluids, which troughs being placed horizontally the plates are perpendicular to the horizon and form so many partitions that are filled with weak nitrous acid and various chemical compounds. Three or more of these troughs, thus composed, are connected by copper wires, and the power of the pile is encreased to a ratio either of the number of the plates or their superficial capacity.[51]

Dinwiddie's notes on the power of the "voltaic battery" clearly reveal that he had access to John Hunter's account of the torpedo and electric eel published in the *Philosophical Transactions* in 1773 and more recent reports in *Nicholson's Journal* in 1803.[52] The result was an accelerating experimental programme in Calcutta dependent on the local making of Galvanic apparatus. While information followed books and journals, the issue soon became the skill of the builders and their assistants.

---

50   DUA, MS 2–726, A 59. Joseph Hume to Dinwiddie, November 25, 1803.
51   Ibid., November 25, 1803.
52   DUA, MS 2–726, D22. Journal of Galvanism, June 24, 1803.

2   Practical Spaces

Dinwiddie tried to turn example into practise. Trial upon trial served to make
his reputation in India among the nabobs, Company agents, and soldiers of the
Crown. He was untidy and promiscuous in his sources, but had learned to be
resourceful in seizing opportunity. By April 1804 he was building more galvanic
batteries, following the models of those in Europe, of the celebrated Humphry
Davy at the new Royal Institution, Martinus van Marum in Haarlem, and W.H.
Pepys in London's Askesian Society. In Calcutta, he built one of 400 plates, and
by July was applying it to local victims of rheumatism like the Calcutta merchant Alexander Colvin. The time had come to prove his skill as instrument
maker. In January 1805 he gave public lectures on Galvanism using a battery
of 300 plates which acted powerfully, and continued throughout the spring to
his subscribers desperate for the latest of European knowledge.[53] Such instruments, of course, revealed the power of Europeans over nature, and undoubtedly confirmed assertions of the force of empire in a way that had failed to
impress in China. There is considerable evidence that he was sufficiently well
versed, and the demand sufficiently high, that Dinwiddie attracted the interest of Alexander Gika, possibly a Romanian Christian living in Calcutta, who
contemplated more than self-application for his own ailments. He also shared
patients with Dinwiddie. Throughout 1806, Dinwiddie finally decided on his
long-delayed return to Britain, and was consequently attempting to sell much
of his scientific apparatus. He was still, however, applying galvanic electricity
to those seeking cures, such as to a Mr. Cracroft, possibly William Cracroft who
had been admitted a student at the College in 1803. There was never any short-
age of frailties nor seductions by promising instruments. He kept at least one
copy of his galvanic battery which he brought with him on his voyage toward
the end of 1806. Stopping at the way station in Cape Town, in January, 1807
Dinwiddie himself became an attraction. Even at the Cape he set up his device
and galvanized several ladies as, he lamely wrote, "it is new here."[54] Unwilling
to miss a chance, Dinwiddie transported the electrical cure to the Cape. We
know that some of these instruments survived his return to London and were
still in the hands of his grandson in 1858 in Liverpool.[55]

---

53   DUA, MS 2–726, D22. Journal of Galvanism, July 28, October 27, 1804; January 19- June 1,
      1805. Cf. Nair, "'Bungalee House Set On Fire By Galvanism,'" 73. Dinwiddie had a list of cures
      by galvanism, including his own spasms, as broad as paralytic affections on the left of the
      head, rheumatism, paralysis of the ankle and foot, night blindness, and St. Vitus dance,
      and tried it in hydrophobia and lockjaw. See DUA, MS 2–726. D22. Loose sheaf at end.
54   DUA, MS 2–726 Journal B 73, January 4, 1807.
55   See DUA, MS 2–726, D 22, in the hand of William J. Proudfoot, October 25, 1858.

Knowledge thus moved across oceans, in the shape of books and bales, of chemicals and experimental things. Yet, Dinwiddie was hardly the only one shipping devices out to India and back. Indeed, by the early nineteenth century it was thought worthwhile enough by the British government to arrange for the East India Company to transport, "freight free", a vast array of philosophical apparatus to the Calcutta Hindoo Sanscrit College. Chemistry and electricity both figured highly.[56] It was obvious, as Karel Davids and Steven Harris have argued, that "long-distance corporations were interested in the collection and transmission of knowledge about all sorts of matters and practices that could keep their operations going and the personnel fit for mission", thereby extending Francis Bacon's early imperial ideal of the "intellectual globe."[57] While exploration was organized by the state, trading houses became the secondary mechanisms of scientific circulation. Hence, Europeans gained an enormous amount of information about the cultivation of plants and practices of production in the Far East, thereby making Canton and Calcutta entrepots of natural knowledge as well as goods and samples.[58] From 1792, Dinwiddie was an incidental player in the Banksian empire.

Dinwiddie played a role in the transit of science ever since he had been recruited in 1795 in Calcutta on instructions of the Board of Trade. He was soon in contact with the little appreciated Dr. Helenus Scott, on the other side of the sub-continent. Scott (1756–1821) was an early member of the Medical Board of Bombay and part of a vast network that included the illustrious and influential Banks, long tied to Henry Dundas at the Board of Trade, as well as privately corresponding with the radical Dr. Thomas Beddoes, and with Dinwiddie himself. Just months before Dinwiddie had set sail with Macartney, Scott had written to Joseph Banks about the purity of alkali that might be used in soap manufactures, followed a few weeks later with an account of alkalis in dyeing.[59] These

---

56 Kapil Raj, *Relocating modern science. Circulation and the Construction of Scientific Knowledge in South Asia and Europe. Seventeenth to Nineteenth Centuries* (Delhi: Permanent Black, 2006), 176–178.

57 Karel Davids, "On Machines, Self-Organization, and the Global Traveling of Knowledge, circa 1500–1900," *Isis* 106 (December, 2015): 866–874, at 869; and Steven J. Harris, "Networks of Travel, Correspondence and Exchange," in Katherine Park and Lorraine Daston, eds., *The Cambridge History of Science*, vol. 3. Early Modern Science (Cambridge: Cambridge University Press, 2006), 341–362, at 345.

58 Raj, *Relocating Modern Science*, 15; Davids, "On Machines, Self-Organization," 871.

59 Helenus Scott to Joseph Banks, January 19, 1792, and February 7, 1792, in Neil Chambers, ed., *The Scientific Correspondence of Sir Joseph Banks, 1765–1820* (London: Pickering & Chatto, 2007), IV, 99–100, 103–105. I wish to thank my former student Ms. Kristen Schranz for drawing my attention to these letters.

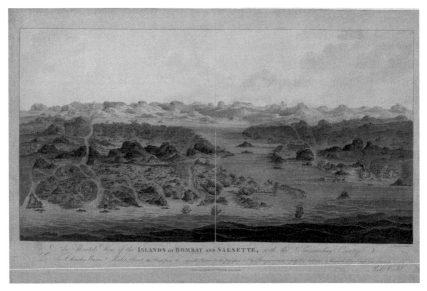

FIGURE 8.6   Salsette Island, Bombay, by J. S. Barth, 1803. Permission of the British library.

intersections illustrate the manner in which natural and mercantile converged with empire, just as Banks and Dundas intended.

Before Dinwiddie appeared in India, Scott had been recognized for "his researches in chemistry and in bringing to light many important discoveries in natural history." He was, by then, engaged in the cultivation of hemp on Salsette Island off the coast which had been occupied by the British since 1774 and controlled by the East India Company since 1782 (figure 8.6). Scott's Baconian projects involved a wide range of substances including sugar, indigo, tobacco, aloes and the much-discussed alkalis. Scott's interests were as wide-ranging as many a late eighteenth-century medic or chemist.[60] For this reason, he attempted to interest Banks in his plans, particularly those relating to dyeing and fixing of colours. He clearly was no slouch, corresponding with the chemist Joseph Black in Edinburgh and was encouraged by Banks in London to look into cochineal and alkali manufactures, as well as various Indian methods of production.[61] And Scott was not alone. For example, William Carey

60   British Library, IOR/H/210. Home Series No. 34. 1792; Anne Bulley, *The Bombay Country Ships 1790–1833* (London and New York: Routledge, 2000), 55, 99, 103, 178. See also Mark Harrison, "Medical experimentation in British India. The case of Dr Helenus Scott," in Hormoz Ebrahimnejad, ed. *The Development of Modern Medicine in Non-Western Countries. Historical Perspectives* (London and New York: Routledge, 2009), 23–41.

61   British Library, Add. MSS. 33979, *Joseph Banks Correspondence*, vol. III, fols. 1–13. Scott to Banks, January 7, 1790; fols. 20–21. April 2, 1790; fols. 127–130. January 19, 1792; fol. 135. March 20, 1792.

was another entrepreneur, botanist and Baptist missionary, among those who produced indigo on the outskirts of Calcutta.[62] With much else, Dinwiddie became curious about the manufacture of manganese, important in experiments in gas chemistry, which he may have learned from Alexander Tilloch's *Philosophical Magazine*.[63] Printed sources sent to India assisted nicely in the kind of commercial intelligence London wished to secure.

Helenus Scott is of particular interest as one of many go-betweens within a wide-reaching imperial network. This was revealed nicely in letters between Scott and Dinwiddie on their discoveries made in India. One special topic of discussion was the subject of nitric acid and medical trials quickly revealed in the *Calcutta Courier* in 1796 and 1797. Dinwiddie seems the most likely source. Scott was not entirely pleased with such a public discussion, as it seemed bound to raise expectations. Such had been the frequent experience in Britain born of too many claims of the utility of experimental philosophy. At the limits of empire this was notably so of laboratory trials not readily replicated. With nitre, however, there was urgency. As he proclaimed to Dinwiddie, "We experience here the happiest effects from it in those venereal Cases where mercury has entirely failed."[64] While there was much concern, there were also many obstacles. Scott and Dinwiddie were pressed to deal with the production of nitric acid through a method employing vitriolic acid, an issue Dinwiddie had attacked by way of a double distillation. Scott wished for a vitriolic acid for his "own use", and not for sale, hoping Dinwiddie knew of a method of "accelerating the Combustion of the Sulfer & the condensation of the Acid..."[65] Vitriolic acid was well known, long used in indigo manufactures. For this reason alone, Scott would have every motive to be interested. It appears Dinwiddie established an

---

62 Raj, *Relocating Modern Science*, 172; Khyati Nagar, "Between Calcutta and Kew: The Divergent Circulation and Production of Hortus Bengalensis and Flora Indica," in Lightman, McOuat, Stewart, eds., *Circulation of Knowledge*, 153–158, 176–178.

63 DUA, MS 2–726, Journal B 72 Calcutta (August 8, 1806).

64 University of Guelph Special Collections, Dinwiddie Correspondence, XS1 MS A164, 94. {improperly catalogued} Helenus Scott to James Dinwiddie, August 25, 1796. Scott corresponded with the radical Thomas Beddoes on the much-debated cures for venereal disease using nitric acid. Thomas Beddoes, M.D., *A Collection of Testimonies Respecting the Treatment of the Venereal Disease by Nitrous Acid, published by Thomas Beddoes, M.D.* (London: Printed for J. Johnson, 1800); See Birmingham Central Library, James Watt Papers 4/23/28. Thomas Beddoes to James Watt, July 7, [1794]; William Blair, *Essays on the Venereal Disease* (London: J. Johnson, 1798), 35, 50ff; [Robert Thornton], *The Philosophy of Medicine: Or, Medical Extracts on the Nature of Health and Disease, Including the Laws of Animal Oeconomy, and the Doctrines of Pneumatic Medicine*, 5 vols.(London: C. Whittingham, 1799–1800); Lance Day and Ian McNeil, eds., *Biographical Dictionary of the History of Technology* (London: Routledge, 1996), 181–2; *A Journal of Natural Philosophy, Chemistry & the Arts*, ed. William Nicholson, 5 (September, 1801): 201–211.

65 Ibid., University of Guelph, XS1 MS A164, 94. August 25, 1796.

acid manufactory with the traders Fairlie, Gilmore & Co, critical among the Bengal merchants, although there was also competition in Calcutta from the firm of Downie & Maitland, financiers and merchants, "in safe packages for carriage."[66] As we shall see, safe transport of chemicals was forever an issue. All of these activities at the edge of empire substantiate the view that science by the early 19th century could not simply rely on philosophical speculation. Science included, indeed depended upon, "instruments, techniques, and services used in the production of knowledge." Teaching and training, as developed in Fort William College, alongside the "improvement of public understanding of knowledge-making activities" tied closely to instruments and chemicals, were thereby primary objectives of trade and empire.[67]

As banal as trade may sometimes seem, like natural knowledge its power lay in erasing the geographic and cultural limits of local spaces. Often the only extant hints may rest in the merchant's balanced books, of hidden requests fulfilled, and even perhaps those not filled—which are, obviously, mostly missing. To follow the trails of goods to India and back might be a dismal task in comparison to the romance of a scientific bounty in a new world. But, of course, some records do turn up and reveal what was once simply the business of private persons at the far poles of a long voyage on East India Company ships. These are the daily traces of a much deeper story even amidst the most curious places. For some, perhaps, this is the attraction of archives otherwise assumed of little apparent value to the history of science. This is the importance, shortly after Dinwiddie's arrival in Calcutta, of an enthusiasm for experiment that defied sea boundaries.

3   The Chemist of Long Acre

We know likewise of extensive correspondence with the chemical manufacturer and trader, Joseph Hume, of 108 Long Acre, in the parish of St. Martin's in the Fields regarding orders for books, apparatus, and glassware.[68] In negotiat-

66   Nair, "'Bungalee House Set on Fire by Galvanism,'" 60–61, & n. 57; see John Mathison & Alexander Way Mayson, comp., *The New Oriental Register and East-India Directory for 1802*...(London: Black's and Parry, 1802); Cf. W.B. O'Shaunessy, M.D., *The Bengal Dispensatory* (Calcutta: W. Thacker & Co., St. Andrew's Library, 1842).

67   Raj, *Relocating Modern Science*, 10.

68   *London Directory* (1794). Hume made an important reputation amongst medical chemists well into the 19th century, most notably for a test for arsenic by nitrate of silver. See Thomas Cooper, *Tracts on Medical Jurisprudence: Including Farr's Elements of Medical Jurisprudence* (Clark, New Jersey: the Lawbook Exchange, 2007), 482; *The Philosophical*

ing the trade in books and chemicals, Hume facilitated the transit of experimental knowledge. A manufacturer in Covent Garden, he was a supplier of
much, from pills and potions, laboratory chemicals, to devices, to accounts of
the latest scientific controversies. But he was no mere conduit. He achieved a
chemical reputation of his own well into the 19th century. Alexander Tilloch's
Philosophical Magazine reveals that Hume was frequently engaged in London's
chemical debates at the turn of the century. Hume was also one of those hybrid
experts whose skill and knowledge crossed the trading zones. His were goods
shipped between practitioner and philosopher evading, if not thereby erasing,
apparent boundaries between daily trials of the laboratory and the seductions
of theoretical speculation.[69] As in books and journals, knowledge travelled in
many forms—in chemical concoctions to cope with diseases of the littoral and
in instruments which, when bought and demonstrated, asserted the virtues
of the European mind. Hume was clearly not alone as instrumental interme-
diary. Shortly after Dinwiddie's arrival in Calcutta, the London instrument
maker William Jones was quick to offer Dinwiddie a planetarium, as he did
to "the first people of science now existing." The most flashy on offer was the
expensive and elaborate Grand Orrery of Lord Bute which, Jones exclaimed in
1795, was "the best instrument ever offered to the public at the price." Devices
like this demonstrated the highest of scientific skill that might serve a lecturer
and amuse an audience.[70] Some devices might be found in shops while others would be created in Calcutta. Manufacturers and traders like Hume were
indispensable in bridging the vast gulf between Long Acre and Dinwiddie, in
Council House Street, off the Bay of Bengal.

One literally explosive difficulty was the shipment of chemicals to India
which he discovered was prohibited by some captains. But Hume found a
way around the problem. Until a reliable supply of acids could be established
in Calcutta, Dinwiddie then had no other means of obtaining the necessary

---

     Magazine, 30 (February-May, 1808): 168–171, 274–280, 356–363; vol. 31 (1808): 161–174; vol.
     40 (July-December, 1812): 105; John Anderson, *A Conspectus of Prescriptions in Medicine,*
     Surgery and Midwifery...Second edition (London: for John Anderson, 1826), 24.

69   Cf. Pamela O. Long, "Trading Zones in Early Modern Europe," *Isis* 106 (December, 2015):
     844. The notion emerges most directly from Peter Galison, *Image and Logic: A Material
     Culture of Microphysics* (Chicago: University of Chicago Press, 1997), 798 and, in turn, fol-
     lowed in Lynn Hunt, *Revolution and Urban Politics in Provincial Societies* (Stanford: Stan-
     ford University Press, 1978).

70   University of Guelph Special Collections, Dinwiddie Correspondence, XS1 MS A164, no.
     66. William Jones to Dinwiddie, March 26, 1795. Dinwiddie was only one of many who
     attracted the offers from Jones in the 1790s. See, for example, William and Samuel Jones to
     Thomas Jefferson, March 9, 1793, National Archives, https://founders.archives.gov/docu-
     ments/Jefferson/01-25-02-0305.

supplies. So, Hume disguised containers of nitrous and vitriolic acids—simply by mislabeling them. In the spring of 1796, he wrote to Dinwiddie that "...you will observe 11 large bottles with Nitric acid, labelled, Spt. Nitricus 9 large bottles Nitrous acid, labelled Stp. Nitrosus—and 3 large bottles of Vitriolic acid, labelled Spt. Sulphuris. I did this lest they might have been examined."[71] An invoice from Hume showed much that an experimental chemist would employ in replicating the latest reports from London, including:

Sulphuric acid–vitriolic ether	*{sulphuric ether}*
Nitrous ether—nitric acid	*{ethyl nitrate}*
Nitric Acid, nine bottles in a case of the same size	
Nitrous Acid, Do. in a case of the same size	
Veget. Alkali-----Mineral alkali	*{alaline salt of sodium}*
Sympathetic Ink—Prussian lime	*{lime water, Prussian blue}*
Prussian Alkali—Muriate of Barytes	*{used to test for sulphuric acid}*
Phosphorous----Tinct. Of Galls	*{rhubarb paper; a test for morphine}*
Caust. Vol. Alkali—Nitrate of Copper	*{copper acetate}*
Acid of Sugar—Corros Sublimate	*{corrosive, mercuric chloride}*
Nitrate of Bismuth—Nitrate of Quicksilver	*{ointment}*
Powder of Manganese—Sulphur	
Colbalt Ore—Bismuth—Zinc	
Crude Sal Ammoniac–Sugar of Lead	*{lead acetate, sweetener, toxic}*
Blue Vitriol—Powder of Luting	*{copper sulphate; fast set luting}*
Mild Vol. Alkali—Litmus–Tumerick	*{used as a rubefacient, re circulation}*

But ship's pursers were not always duped. Hume was soon frustrated in efforts to find a compliant purser or captain. In 1796 he told Dinwiddie,

> Every thing was ready months ago, but I was obliged to unpack them on account of the mineral acids being prohibited; however I have found means, by changing the names, to get above 150 lb[?] of nitrous and nitric acid, and a few pounds of vitriolic acid. Had I known the trouble and anxiety of mind the executing of your orders would have occasioned I certainly would never have engaged in it ...[72]

---

71 DUA, MS 2–726. Dinwiddie Correspondence, A 75. Jos. Hume to Dinwiddie, dated London, May 17, 1796.

72 DUA, MS 2–726. Dinwiddie Correspondence, A 75. Jos. Hume to Dinwiddie, dated London, May 26, 1796.

Henceforth, he seems to have enlisted the firm of David Scott & Co closely tied to the East India Company. By this means Dinwiddie received his books, and the fragile glassware and rather unstable chemicals some of which unfortunately spilled or evaporated in transit.[73] It might be noticed that these substances were all being sought just as Dinwiddie and Helenus Scott exchanged their views on nitric acid in the treatment of venereal disease, then a topic of intense concern. Dinwiddie, as we now know, was driven by the much-contested pneumatic chemistry and this is likely one of the reasons he sought manganese from Britain, as it was used often in the production of factitious airs.[74] Hume proved an excellent source of chemicals, as he manufactured many sold from his shop in Long Acre.[75] But he also provided not only copies of journals, by Nicholson and Tilloch, and of the *Encyclopaedia Britannica* which all discussed the latest in European chemistry. At the very least, they could be used to pack the crates of chemicals.

Hume knew exactly the difficulties an experimenter might face, notably more so from reports of the failures of some of the chemicals to arrive intact. In 1800 he wrote to Dinwiddie about the refusal of Captain Price of the ship Lord Hawkesbury to assist and, he complained,

> ...of the disastrous fate of the nitrous acid. I very much regretted this loss, as I had taken great pains to make the acid strong and pure; and though it might have...[unintellible] the mouths, though well stopped with glass, yet I could defy any other accident to happen. When this or any other liquid is sent from a cold climate an expansion will naturally take place, and especially in this acid, which is apt to generate a nitrous gas. I cannot account for the oxidation of the phosphorous—what I sent was perfectly pure though made from bones and the price you mentioned to have had from Godfrey's was certainly made from inspissated urine. I have never tried it, but would not Phosphorous keep better in alcohol? One thing I must observe to you, that Phosphorous in water should be kept in a dark place, for the light assists it to decompose in water. However the phosphorus may be oxidated it cannot be lost; for, I need not tell you, it must have gained something by the change you complain of. In respect to the

---

73   DUA, MS 2–726. Dinwiddie Correspondence, A 75. Jos. Hume to Dinwiddie, dated London, May 27, 1796.

74   DUA, MS 2–726. Dinwiddie Correspondence, A 75. Jos. Hume to Dinwiddie, London, Apr. 17, 1797.

75   In 1799 the sale of crystals of soda could be obtained from the Alkali Manufactory, North-street, Poplar, London. Hume was evidently one of a number of retailers. See Arthur Young, *Annals of Agriculture, and Other Useful Arts*, 32 (Bury St. Edmunds, for the editor, 1799): 434.

investment of the glass I can say nothing about its being damaged, for in that, and the circulating machines with air pump receiver &c. I had nothing to do in packing them...

Hume added further, given Calcutta efforts with indigo dyes,

On many accounts I do not think the oxy-muriatic acid a fair test—probably its power (I mean the cochineal) in dying a specific quantity of cloth, and comparing this for colour and quantity with what has been dyed by the Spanish, would be a better and less objectionable mode. There are many things which I think might be accomplished in India such as making of phosphorous, the nitrous, sulphuric & muriatic acids &c. &c. Surely the bones from of Seringapatam would make phosphorous by proper calcination and decomposition with sulphuric acid. I suppose you are laughing at all this philosophic speculation.[76]

Hume's interests were not simply in marketing his chemicals, as he was exceedingly knowledgeable and could supply chemists in India with the latest printed sources. By 1802 he was advising Dinwiddie about the best use of "such cylinders as we try experiments on the specific gravity of liquids by means of the Hydrometer; and these will prove very useful in making a great number of expts. in pneumatic chemistry, and in eudiometry &c. &c." Pneumatic chemistry could be advanced by galvanic experiments even if sometimes hindered by monsoon damps. And in 1803 they were corresponding about galvanic apparatus, one purpose of which was the production of airs which could then be used in various pneumatic trials.[77]

4   Electrical Airs

Dinwiddie put such intelligence to good use in Calcutta. In May of 1802 he was noticing oxygen produced by application of the galvanic battery, and of alkali produced by soaking the plates in muriate of ammonia which decomposed in the electrical current.[78] Over the next two years, as he received more

---

76   DUA, MS 2–726. Dinwiddie Correspondence, A 75. Hume to Dinwiddie, dated Long Acre, London, Aug 29, 1800.

77   DUA, MS 2–726. Dinwiddie Correspondence, A 75. Hume to Dinwiddie, dated Long Acre, July 8, 1802 and November 25, 1803.

78   DUA, MS 2–726. D 22 Journal of Galvanism (Calcutta), no. 20 (May 1, 1802), and no. 23 (March 5, 1803).

books and chemicals from Hume and tried tapping the resources of Scottish merchants in Calcutta, his galvanic and pneumatic experiments blossomed. In 1804 he commented upon the reports of Martinus van Marum in Haarlem and of Humphry Davy in London's new centre of experimental chemistry at the Royal Institution. Chemical intelligence had crossed great spaces. By March and April he was trying Davy's preference for nitric acid.[79] That was hardly the end of the experiments, as we now know. Whatever he gleaned from London journals induced a farther range of trials. He was ever experimenting with different sizes of batteries and different metals, ultimately employing sal ammoniac [ammonium chloride] solution and diluted muriatic acid [corrosive hydrochloric acid].[80] While Davy generated a celebrated reputation and married wealth in London, Dinwiddie quietly and anonymously manufactured his fortune in Calcutta.[81]

There was knowledge in scientific goods. If the nature of global trade was the expansion of power, whether of the British state or of the East India Company, it induced also an exchange of ideas focussed obviously in the construction and purpose of scientific instruments by which the world would be explored. Parker's burning lenses, the dismantling of which so agitated Dinwiddie in China, could potentially be used in the melting of metals or reducing ores even for manufacturing. Likewise, pneumatic productions of new airs, or electrical generators and batteries could be of use both in chemical experiments and medical application, by a Dinwiddie or a Davy among many. Similarly, Dinwiddie not only transported tea plants from China for their cultivation in Calcutta, he was employed in seeking out indigenous knowledge. Hence indigo, hemp, soap, flax, tobacco, alkalis, salts and nitric acid were of potential worth in a wealth of manufactures and Dinwiddie explored them all. If there were discoveries in the East they need close examination. Manufactures were not the only objective.

Dinwiddie existed on the periphery of power, whether social or imperial. His real interest was in demonstrating knowledge to the curious and in the end, by force of circumstance, coping with the great divide from Europe. By the summer of 1804, his own devices were so much in demand that he had sold 5 sets of the galvanic apparatus "with many orders on hand."[82] But he did not leave

---

79   DUA, MS 2–726. D 22 Journal of Galvanism (Calcutta), no. 38 (March 4,1804) and no. 39 (April 6, 1804).

80   DUA, MS 2–726. D 22 Journal of Galvanism (Calcutta), (October 28, 1803) and (January 1 and 19, 1805).

81   Jan Golinski, *The Experimental Self. Humphry Davy and the Making of a Man of Science* (Chicago and London: University of Chicago Press, 2016).

82   MS 2–726, D 22 Journal of Galvanism, no. 43, July 16, 1804. In the next year Dinwiddie received an enquiry from a Captain Price, of the 9th Regiment, Madras Infantry. It is

in India one of his best galvanic devices which had continually attracted the attention of the physically infirm. To the last, even as he disposed of devices, Dinwiddie had an addiction. His last voyage to London proved only a pause. His accumulation of instruments revived virtually as soon as he arrived on London docks—induced, undoubtedly, by encountering the many showmen still competing in the metropolis. The world with which he had long corresponded, he now re-entered. Almost immediately he met Alexander Tilloch of the *Philosophical Magazine* as we might have expected, soon of course his ally Hume, his own reputation such that he was invited to join the Royal Institution where Davy performed.[83] With the crossing back, the barrier of vast distance had been erased.

When Dinwiddie ultimately landed London in 1807, no ideologue of empire, slightly infirm and his hearing fading, he settled into a world that was in the middle of another conflict with the French. His reception was one of a genteel respect, a man of earned wealth and experimental skill, known to the great men around Joseph Banks, and soon meeting often with the likes of the astronomer William Herschel at Slough with whom he discusses telescopic improvements (figure 8.7).[84]

In London, he obsessively made the rounds of the clubs and societies where tradesmen and gentlemen alike met the promise of scientific discovery. It is as though he wanted to miss no scientific news, to make up time and to remake connection. Many metropolitan groups had by then their own cabinets of apparatus, their fellows no longer quite so restricted by social origins which, as Dinwiddie knew above all, were hardly natural and entirely incidental although, unlike some, he never dared to reach the republican conclusions Burke so much feared.

He restored his links to London instrument makers, notably with William and Samuel Jones who still traded in Benjamin Martin's orreries, and with Hume of Long Acre who had been his source for acids while in Calcutta. He thought seriously about giving new lectures, notably on Galvanic experiments. But he also found there was much more afoot than when he had last seen London and would have encountered vast competition despite his wealth of experience in public performance. This drew him to the orbits of the Askesian

---

unlikely to have been the same Captain Price, captain of the East Indiaman, *Lord Hawkesbury*, which carried some of Dinwiddie's chemical goods from Hume in London. See, DUA MS 2–726, Correspondence, 2_2. Captain Price to Dinwiddie, February 27, 1805.

83   MS 2–726 Journal B-73. May 8 and May 16 & 19, 1807.

84   DUA, MS 2–726, Journal B-74, Oct. 21, 1807; Journal B-77, Mar. 20, Apr. 10, 13; Cf. Schaffer, "Instruments as Cargo," 236.

FIGURE 8.7   William Herschel, 1790s, photogravure after Lemuel Abbot,
                Permission of the Science picture library.

Society, to the audience of the City Philosophical Society where he ridiculed
the cockney style of John Tatum, and to lectures put on by the Spitalfields
Mathematical Society in East London. Most significantly, of course, was his
attendance at the demonstrations at the Royal Institution, notably Davy's dramatic lectures especially on nitrous oxide and electrolysis. Dinwiddie was ultimately placed on its Committee of Chemistry, Geology and Mineralogy.

Even while overshadowed, Dinwiddie still attended Davy's spectacular displays. These confirmed the power of instruments to conquer the philosophical
world, even if deeply riven by the smug disputes of Davy's jealous rivals in the
Royal Society. There were precisely refined empires in London too. Even so,
the great imperial reach to China and India ultimately defied social, political,

and intellectual barriers, aside from bitterly-contested boundaries of the bull-pits of London's powerful gentlemen's clubs. Instruments, like Davy's batteries, clearly made great reputations possible if seemingly less so in the distant Far East than in the centre of imperial London. Some begrudgingly noticed a new world had formed from the crucible of experimental experience, through machines to measure geographies and test nature's forces rather than merely to provide amusement to jaded emperors. By the end of the 18th century, ideas had crossed vast seas in many a vessel. As much as celebrated voyages, devices made discovery possible across the intellectual globe. They were powerful agents of knowledge when an experimental enlightenment continually redefined its purpose from Kew and Fleet Street to the Far East and back.

CHAPTER 9

"Both by Sea and Land": William Whiston, Longitude, and the Measurement of Space

Simon Werrett

Since the publication of Dava Sobel's popular book *Longitude* in 1995, there has been much debate on the eighteenth-century quest for a method to determine longitude at sea, for which the British government offered substantial rewards.[1] This discussion has transformed the picture of the longitude story which Sobel presented, shifting from an account centred on the struggling lone genius of clockmaker John Harrison to a story of collaborative labours among a range of artisans, administrators, astronomers, and navigators deploying a variety of methods to solve the problem. But while there has been much progress in the history of longitude, this work has still remained largely focused on the issue of longitude at sea. The typical trajectory encountered in histories of longitude moves back and forth between the workshops of artisans such as Harrison and authorities at the Admiralty or Greenwich Observatory to the Atlantic and Pacific oceans during trials of astronomical and chronometric solutions. Not all successful longitude solutions ended up as maritime ventures, however. This essay offers an alternative trajectory by exploring another longitude solution which failed at sea but rather prospered as a means of finding locations on land. This was the scheme of William Whiston and Humphry Ditton to use rocket signals to discover the longitude.

---

1   Dava Sobel, *Longitude: The True Story of a Lone Genius Who Solved the Greatest Scientific Problem of His Time* (New York: Penguin, 1995); William J.H. Andrewes, ed., *The Quest for Longitude* (Cambridge, MA: Collection of Historical Scientific Instruments, 1996); Derek Howse, *Greenwich Time and the Longitude* (London: Philip Wilson, 1997); J.A. Bennett, "The Travels and Trials of Mr Harrison's Timekeeper" in Marie-Noëlle Bourguet, Christian Licoppe & H. Otto Sibum, eds., *Instruments, Travel and Science: Itineraries of Precision from the Seventeenth to the Twentieth Century* (London: Routledge, 2002), 75–95; Katy Barrett, "'Explaining' themselves: The Barrington papers, the board of longitude, and the fate of John Harrison," *Notes and Records of the Royal Society* 65 (2011): 145–162; Richard Dunn, Rebekah Higgitt, *Ships, Clocks and Stars: The Quest for Longitude* (Glasgow: HarperCollins, 2014); Simon Schaffer, "Chronometers, charts, charisma: on histories of longitude," *Science Museum Group Journal*, 2 (Autumn 2014) DOI: http://dx.doi.org/10.15180/140203

The plan, first put forward in 1713, is best known today as a prelude to the Longitude Act of 1714, and as a failure, lampooned by the Scriblerus Club and William Hogarth as the impracticable work of a madman.[2] But tracing the development and fate of Whiston and Ditton's methods shows that they endured beyond the era of the Longitude act and helped to shape land measurement, surveying and geodesic practices on numerous occasions in the eighteenth and early nineteenth centuries. The essay also pays attention to the geography of the scheme itself. While the project was decried for its impracticability at sea, it was never actually tried there. Rather, the fate of the project depended on a variety of trials on land. The idea of using rockets owed much to landed interests and the royal and state fireworks spectacles of early eighteenth-century London, while the political and religious controversy associated with fireworks helped to direct the trajectory of rockets as solutions to questions of position-finding.

1   **Whiston and Ditton's Projects**

In July 1713, a letter appeared in the *Guardian* from the natural philosophers and mathematicians William Whiston and Humphrey Ditton announcing that they had "a Discovery of considerable Importance to communicate to the Publick" and promoting the value of finding a means to measure longitude at sea.[3] The plan was explained further in a short book published in the summer of 1714, *A New Method for Discovering the Longitude Both by Sea and Land*. It began by discussing the firing of shells and rockets at sea, to serve as a means to find the distance from known points.[4] According to the plan, ships were to be anchored at east-west stations along the trade routes and at midnight (Peak of Tenerife time, 30 degrees west of the London meridian) they would fire a shell or rocket to a fixed altitude, which would explode in a bright "ball of fire." Using this fireball as a signal, ships sailing in the area would take a compass reading of the direction of the signal ship and calculate their distance from it. The difference in time between the flash of the explosion and the sound of it could be measured and used to find the distance between the station and

---

2   See e.g. Sobel, *Longitude*, 50; Howse, *Greenwich Time*, 55–58; Dunn, Higgitt, *Ships, Clocks and Stars*, 39–42, follows the account of Whiston and Ditton in Simon Werrett, *Fireworks: Pyrotechnic Arts and Sciences in European History* (Chicago: University of Chicago Press, 2010).

3   *Guardian*, July 14, 1713 (no. 107), n.p.

4   William Whiston and Humphrey Ditton, and *A New Method for Discovering the Longitude Both by Sea and Land* (London, 1714). A second, revised edition appeared in 1715.

the ship, or the length of time the projectile was visible in the air. The angle of the projectile above the horizon could also yield the longitude. The shells would ideally explode at 6,440 feet because this would make them visible for 100 miles around. Such a scheme, if put into practice, would "tend to the common Benefit of Mankind" and the "Honour and Advantage of this our Native Country."[5]

Larry Stewart has located the Whiston-Ditton scheme within the growth of a Whig-centred public science exemplified by the projecting spirit of the "longitudinarians" in the first decades of the eighteenth century.[6] Historians have typically discussed the scheme as a prelude to the Longitude Act of 1714. From 1713 Whiston and Ditton promoted their rocket signals alongside efforts to lobby for a public reward should a solution to longitude at sea be found. Petitions in April and May of 1714 resulted in the creation of a Parliamentary committee to consider them, and by early July the Longitude Act was passed.

While Whiston and Ditton were instrumental in seeing the act through, however, they failed to win any reward. After Ditton died in October 1715, Whiston pursued the project alone, but by 1719 he had abandoned it, turning instead to magnetic variation as a possible solution to the longitude. In the meantime, Whiston's project was subjected to critique and ridicule. Isaac Newton cautioned that the scheme was among "several Projects, true in the Theory, but difficult to execute" though he also thought the same of using precision watches.[7] By the 1730s, the scheme had been widely lampooned.[8] In 1735 William Hogarth famously depicted Whiston's project as the raving of a madman, scrawled on the wall of Bedlam lunatic asylum in the final plate of *A Rake's Progress*.[9] Subsequent assessments continued to view the scheme as folly, foolishness, or misguided projecting. In 1750, James Kirkpatrick lamented the quest for longitude in his poem *The Sea-Piece: A Narrative, Philosophical and Descriptive Poem*.

> O! Might some grand Discov'rer plainly show,
> With like Facility, the Length we go,
> What future Crouds beneath the Glebe may sleep,
> Who else shall perish in the pathless Deep!

---

5   Whiston and Ditton, *A New Method* (1714), 27.

6   Larry Stewart, *The Rise of Public Science: Rhetoric, Technology, and Natural Philosophy in Newtonian Britain, 1660–1750* (Cambridge: Cambridge University Press, 1992), 183–212.

7   Newton, quoted in Howse, *Greenwich Time and the Longitude*, 57.

8   See e.g. the "Ode, for Musick, on the Longitude," by Jonathan Swift in Swift, Pope, Gay and Arbuthnot's *Miscellanies. The Last Volume* (London, 1727), 172–3.

9   For interpretation, see Jenny Uglow, *William Hogarth: A Life and a World* (London: Faber, 2011), 440–442.

But this not Whiston's Rockets might attain,
And haply Newton thought the Search was vain.
Knowledge with Ign'rance mixt, and Joy with Woe,
By Fits we're happy and by Halves we know.[10]

The scheme's image was thus secured as a failed project for determining longitude at sea, a place it has occupied in historiography of the longitude ever since. But the land mattered to Whiston and Ditton's project just as much as the sea. Their scheme depended on land-locked resources to create the method and the grounds on which it was attacked were as much to do with the territory of religion and Whiston's trespasses into dangerous areas of belief as with impracticability at sea. Even then, for the remainder of the eighteenth century the rocket scheme was taken up as a more promising method for cartography, surveying and geodesy, using British landscape and topography to succeed.

Larry Stewart has proposed that William Derham's experiments of 1704 to measure the speed of sound were a significant inspiration for Whiston and Ditton's project.[11] London's ecclesiastical geography shaped Derham's trials providing resources for surveillance. Derham had guns fired at a prearranged time and listened for the sound at a distance of up to twelve and a half miles, measuring the delay to derive the speed.[12] Tall London churches, and no doubt networks of willing ministers, provided the listening locations, in places such as Dagenham, Rainham, Barking and Blackheath.[13] Derham's own residence at Upminster church, Essex, featured a platform erected on the church spire for making observations.[14]

London's royal fireworks spectacles on the Thames were also an important resource for the Whiston-Ditton scheme. In the introduction to *A New Method*, Whiston explained that the idea of using ordnance to create long-distance signals had occurred to him after hearing the thundering sound of guns from far-distant engagements with French and Dutch ships in the previous century.[15]

---

10   James Kirkpatrick, *The Sea-Piece: A Narrative, Philosophical and Descriptive Poem* (London, 1750), 32–33.

11   Stewart, *Rise of Public Science*, 186.

12   A.D. Atkinson, "William Derham, F.R.S. (1657–1735)," in *Annals of Science* 8 (1952): 368–392, on 381–2.

13   William Derham, "Experimenta & Observationes de Soni Motu, Aliisque ad id Attinentibus, Factae a Reverendo D. W. Derham Ecclesiae Upminsteriensis Rectore, & Societatis Regalis Londinensis Socio," *Philosophical Transactions* 26 (1708 - 1709): 2–35, on 13.

14   A.D. Atkinson, "William Derham, F.R.S. (1657–1735)," *Annals of Science* 8 (1952): 368–392, on 374–5.

15   Whiston and Ditton, *A New Method* (1714), 19.

But Whiston and Ditton soon became concerned that such sounds might not travel over long distances at sea. Whiston claimed that attending a fireworks display on the Thames in the Spring of 1713 reignited their enthusiasm,

> A day of extraordinary Fire-works happen'd (it was the Thanksgiving day for the Peace, July 7th, 1713.) the Contemplation of which, did much revive and encourage this Notion: and the certain Account he [Whiston] soon had, that those Fire-works, nay, the small Stars, into which the Rockets commonly resolv'd themselves, were plainly visible no less than 20 Miles, put an end of his doubts immediately; and made him very secure, that such large Shells as might be fir'd at a vastly greater height, would for certain be visible for about 100 Miles; which he look'd on as nearly the limit of Sounds also, as to any purposes of Longitude.[16]

The fireworks marked the peace of Utrecht of April 1713, concluding the War of the Spanish Succession. The court artist Sir James Thornhill designed a temporary edifice or "machine" from which the fireworks would be played off on the Thames, with two pairs of coupled columns supporting allegorical figures and a victory wreath enclosing Queen Anne's monogram surmounted by a crown. The machine was floated on rafts on the river, while gunners Henry Hopkey and Albert Borgard oversaw the pyrotechnics. These included "1500 great and small water rockets" and "21 standing Rockets."[17] Spectators watched from rafts and presumably included Whiston who was inspired by them to restart the longitude scheme. He went on to propose the use of mortar-propelled shells creating "Fire or Light" for the scheme, or rockets whose ascent could be carefully managed and whose garniture of stars made a brilliant light in the sky when they exploded.

It might seem obvious today to use rockets as signals, but this was not the case in Whiston's time. Natural philosophers had been interested in ordnance and fireworks for many decades, but had only recently begun to treat them as experimental objects. As recently as 1696, the mathematician and weaver Robert Anderson had written in *The Making of Rockets* that,

---

16   Whiston and Ditton, *A New Method* (1714), 23.

17   There were, "1500 great and small water Rockets; 5 large water Pyramids; 4 water fountains; 13 Pumps; 21 standing Rockets, with lights all swimming on the Water; 84 of coll. Borgard's large and small Bees' swarms, half of which were set with lights to swim on the Water." Sir James Thornhill, *Exact Draught of the* firework *that was performed on the River Thames, July 7th, 1713, being the Thanksgiving day for the Peace Obtain'd by the Best of Queens*, British Museum, Crace collection of prints, portfolio V, sheet 29.

in the many Volumns great and small that I have read relating to [rockets] in our own and other Languages; I do not find the least Pretence or Thought of doing that which here is undertaken, viz. To raise so great a heap of Fire, and to confirm the Fact by the greatest Proofs that can be had or wish'd for, which are Experiments and Demonstrations Mathematical.[18]

Anderson was one of the first people to investigate rockets experimentally. In his book, he determined appropriate charges of gunpowder for different sizes of rocket, making experiments in Wimbledon to measure their trajectories. Measuring the horizontal range of the rocket, he constructed a parabola to model its trajectory then calculated the height of the apex of the parabola. He concluded a six-inch diameter (30 lb) rocket would rise to "three hundred eighty and eight Yards, [and] had it been [fired] Perpendicular, its height would have been four hundred and five Yards."[19] Whiston may have read this work, but he did not cite it in the *New Method.* He did, however, cite another book by Anderson as evidence of the height to which mortar shells might be fired.[20]

Just as the local geography of pyrotechnic trials fed into the Whiston-Ditton scheme, so did the local politics surrounding their use. To make experiments with rockets was quite unusual. In late seventeenth-century London, fireworks were a subject of controversy, used less as benign entertainments and more as tools of political and religious propaganda. Courts and cities across Europe staged elaborate and expensive fireworks displays to celebrate the crown and state. The 1713 display was a typical performance in combining temporary festive architecture with "artificial fireworks" such as rockets, wheels, fountains, suns and stars. In London, these events were highly charged affairs. By the 1660s the anniversary of the failed Gunpowder Plot was celebrated with anti-papist pyrotechnics, when "many bonfires are lit all over the town in celebration, and a great lot of fireworks are let off and thrown amongst the people."[21] Throughout the late seventeenth century, fireworks were used to promote or attack different religious and political positions. Many in London viewed fireworks as nothing less than tools of incendiary enthusiasm and Papism. Enthusiasm was itself routinely associated with an "inflamed" or "fiery" temperament.

---

18   Robert Anderson, *The Making of Rockets in Two Parts, the First Containing the Making of Rockets for the Meanest Capacity, the Other to Make Rockets by a Duplicate Proposition, to 1000 Pound Weight or Higher Experimentally and Mathematically Demonstrated* (London, 1696), epistle dedicatory.

19   Anderson, *The Making of Rockets*, 40–42.

20   Whiston and Ditton, *A New Method* (1715), 38.

21   *The Journal of William Shellinks' Travels in England, 1661–1663,* trans. and ed. Maurice Exwood and H.L. Lehrmann (London: Royal Historical Society, 1993), 172.

Supposed Jesuit acts of pyrotechnic terror were notorious. *Pyrotechnica Loyolana* published in 1667, accused the Jesuits of starting the great fire of London using fireworks. Londoners fought fire with fire.[22] The 1670s witnessed the famous London "pope-burnings" stoked up by opponents of the Duke of York's refusal to renounce Catholicism.[23] An effigy of the pope was processed through the city then burned at Cheapside, an event revived in the years 1711 to 1714 to express Whig disdain for the Tories.[24]

Such was the backdrop for Whiston's proposals to use rockets, which might easily appear as another enthusiastic scheme involving pyrotechnics. Whiston was of course notorious for his radical religious views, as a heretical believer in the Arian doctrine denying the Trinity.[25] From 1701 Whiston enjoyed the patronage of the secret Arian Isaac Newton, who arranged Whiston's succession to the Cambridge Lucasian chair of Mathematics. At Cambridge, Whiston became convinced of the Arian or Socinian doctrine, coming to believe that the original primitive Christianity had been sullied by the early Church councils' rejection of the claim of the Alexandrian presbyter Arius that Christ was a subordinate entity to God. Whiston, like Newton, was convinced that the Church had clung to a profoundly erroneous doctrine for hundreds of years which needed to be abolished to restore true religion.

Such views were controversial and even dangerous, and friends and fellows urged Whiston to desist from publicly declaring his beliefs. Whiston, however, insisted on publishing anti-Trinitarian views, leading to his dismissal from Cambridge in 1710. He moved to London to begin a new career in lecturing, publishing and projecting, often working with the nonconformist preacher and Newton's protegé, Humphrey Ditton, a lecturer in mathematics at Christ's Hospital. Controversy surrounding Whiston did not then die down and he continued to experience repeated condemnations for heresy during the years he worked on the longitude scheme.[26]

---

22   Werrett, *Fireworks*, 78–79.

23   O.W. Furley, "The Pope-Burning Processions of the Late Seventeenth Century," *History* 44 (1959): 16–23.

24   *An Account of the Mock Procession of Burning the Pope and the Chevalier de St. George: Intended to be Perform'd on the 17th Instant, Being the Anniversary of Queen Elizabeth of Pious and Glorious Memory* (London, 1711); See also the account of November 1714 in *The Spectator*, ed. Donald F. Bond, 5 vols. (Oxford: Clarendon Press, 1987), vol. 5, 106.

25   Whiston's religion is explored in James E. Force, *William Whiston: Honest Newtonian* (Cambridge: Cambridge University Press, 1985); Steven D. Snobelen, "William Whiston: Natural philosopher, prophet, primitive Christian" Ph.D thesis, University of Cambridge, 2000.

26   William Whiston, *Memoirs of the Life and Writings of Mr William Whiston: Containing Memoirs of Several of his Friends Also*, 2 vols., second edition (London, 1753), vol. 1.

Critics of Whiston and Ditton's longitude scheme rapidly picked up on Whiston's heretical beliefs.[27] Their distaste for his Arianism, not to mention the controversial nature of pyrotechnics, undoubtedly contributed to the scepticism and criticism which greeted the longitude scheme. Criticisms of the longitude scheme were often interspersed with scoffs that Whiston was a fiery enthusiast. He was the "grand ignis-fatuis of London", and "a false Fire who shakes the very Foundations of the Christian Faith."[28] Whiston's allies used similar language in his defence. John Harris, author of *Lexicon technicum*, attacked the Tory preacher Henry Sachaverell after his Gunpowder Day sermon of 1709 condemned nonconformists and dissenters of any sort: "The Heat and Fire of your Spirit and Temper makes you, as it doth all Men, ignorant and precipitant in your Judgments of things; your fiery Zeal is without knowledge." Whiston provided ammunition for his enemies, by stating in the *New Method* that if the longitude scheme worked it would allow "the Propagation of our Holy Religion, in its original Purity, throughout the World", no doubt a reference to Arianism. Critics quickly pounced on these associations.

> I do not know... how well our Will-with-a-Wisp understands *Longitude* after all; and yet I verily believe... he is a profess'd *Latitudinarian* in an Ecclesiastical Sense, to all Intents and Purposes of... Fanaticism, against the establish'd... Church of England.[29]

The Tory Scriblerians condemned the scheme. John Arbuthnot said it was the "most ridiculous thing that was ever thought on."[30] Even Whiston's sympathizers were troubled. Newton disowned Whiston for being too much of a controversialist.[31] Francis Hare, warning of the dangers of straying from orthodox Christianity, wrote of Whiston in 1714, "'Tis the poor Man's misfortune to have a Warm Head and to be very zealous in what he thinks the cause of God... They

27   This issue is explored in Werrett, *Fireworks*, 96–98.

28   N.N., Gentleman formerly of Queen's College Oxon., *Will-With-a-Wisp; Or, the Grand Ignis Fatuus of London. Being a Lay-Man's Letter to a Country-Gentleman, Concerning the Articles Lately Exhibited* (London, 1714), 'Advertisement'.

29   N.N., *Will-With-a-Wisp*, 59, 61. Cf. Stephen Snobelen and Larry Stewart, "Making Newton Easy: William Whiston in Cambridge and London," in Kevin C. Knox and Richard Noakes, eds., *From Newton to Hawking* (Cambridge: Cambridge University Press, 2003), 135–170.

30   John Arbuthnot to Jonathan Swift, 17 July 1714, in Jonathan Swift, *The Works of Jonathan Swift*, ed. Thomas Roscoe, 2 vols. (London, 1843), vol. 2, 510–511 on 511; see also Force, *William Whiston: Honest Newtonian*, 25–26.

31   On Whiston's relationship with Newton, see Stephen D. Snobelen, "William Whiston, Isaac Newton and the crisis of publicity," *Studies in History and Philosophy of Science* 35 (2004): 573–603.

that speak most favourably, look upon him as craz'd, and little better than a *Madman.*"[32] Here then was the origin of the association of Whiston's scheme with madness. But it had as much to do the religious scandal that surrounded him as any impracticability at sea.

2 Longitude on Land

Certainly nothing came of Whiston and Ditton's scheme for finding longitude at sea. But it was not the case that contemporaries universally viewed the method as impractical. Newton was sceptical, but he was also sceptical of chronometers. Others evidently thought the method good for use on land, if not at sea. Indeed, Whiston and Ditton's book was about more than just solving the longitude at sea. The title was, after all, *A New Method for Discovering the Longitude Both by Sea and Land.* In section eight of the expanded 1715 edition, the authors explained that their method would be useful for land surveying:

> If in any clear and calm Night a sufficient Number of such Explosions were made at proper Distances in any Country, and convey'd in order from one to another; so that the Second Mortar or Gun were fired when the Light of the First was seen, or otherwise; with the Observation of the exact apparent Times when they were made, when the Light was seen, what Angle, or how long that Light was above the Horizon, and what Azimuth it had; both the Longitude and Latitude, as well as the Distance and Position of all these Places, might by this means be readily determin'd at Land; especially if the Experiments were repeated several times, and were compar'd one with another. And by the same Observations every where, the Longitude and Latitude, Distance and Position, of all other Neighbouring Places from those, and so from another, might be readily determin'd also.[33]

The authors added another use for rockets, as communication signals.

> Signals of all Sorts may be given by this Method, by mutual Agreement... no one knows how far this Method of Communication... may be improv'd; and how great a Convenience may hence arise to the Several

---

32   Francis Hare, *The Difficulties and Discouragements Which Attend the Study of the Scriptures in the Way of Private Judgment* (Boston, 1749), 22.

33   Whiston and Ditton, *A New Method* (1715), 85–86.

Parts of the Globe; especially in the Way of Trade and Commerce; and even for the Propagation of Knowledge both Divine and Human throughout the World.[34]

The proposed methods, which the authors went on to elaborate in more detail, in fact inspired a variety of uses for rockets in land surveys and as signals for military communications in subsequent decades. By the 1820s, observations of rocket explosions following methods with roots in the Whiston-Ditton scheme became a highly-regarded means to find time differences between two locations.

The period 1700 to 1850 has been called the "golden age of the local land surveyor", a period when medieval written surveys were replaced by cartographic surveys conducted to meet growing demand from land owners interested in agricultural or manufacturing improvements or in selling or leasing parts of their estates.[35] Larger-scale maps also multiplied in the eighteenth century, serving as enlightened decorations, tools of state-craft or simply as the most accessible form of geographical information for a growing class of consumers.[36] Whiston joined an expanding cadre of map-makers and surveyors in plying this trade, catering for a diverse audience ranging from landed gentry to country parsons.

In January 1715, as Stewart has shown, following efforts to establish the Parliamentary reward for finding longitude at sea, Whiston, along with Ditton's widow, published a proposal to produce by subscription a new set of maps of England and Wales made "according to Mr. Whiston's and Mr. Ditton's New Method for the Discovery of the Longitude."[37] The proposal indicated that once five hundred sets of maps were subscribed to, the project would take to the field, measuring meridians at Rye, London, Chichester, Salisbury, Bridport and Dartmouth, "at every Degree of Longitude from the Royal Observatory at Greenwich." These would be traced by "the Casting up of Rockets, or Balls of Fire", as would a series of parallels across the country.[38] If a thousand sets of maps were subscribed to, a mortar would be purchased, capable of shooting

---

34   Whiston and Ditton, *A New Method* (1715), 84–85.

35   Alan R. H. Baker, *Studies of Field Systems in the British Isles* (Cambridge: Cambridge University Press, 1973), 10–12.

36   Mary Sponberg Pedley, *The Commerce of Cartography: Making and Marketing Maps in Eighteenth-Century France and England* (Chicago: University of Chicago Press, 2005).

37   [William Whiston], *Proposals (by Way of Subscription) For a New and most Correct Sett of Maps for England and Wales, By an Actual Survey of the same, according to Mr. Whiston's and Mr. Ditton's New Method for the Discovery of the Longitude* (London, 1715); Stewart, *Rise of Public Science*, 189–90.

38   [Whiston], *Proposals (by Way of Subscription) For a New and most Correct Sett of Maps*, 1.

a ball of fire a mile high. This, Whiston reckoned, would be audible 20 miles away and visible 50 to 60 miles away. The mortar would be fired at the intersection of meridians and parallels, when designated persons in towns nearby would take the bearing of the explosion, "and by the Interval of the Sound, the Distance of the Explosion from the Church, or most Central Part of the Town." The speed of sound was to be taken as William Derham's measure of eight miles in 37 seconds. Trained observers from London would be sent to each market town. A large musket would be used to make signals for further measurements of the locations of landmarks close to market towns. A table was provided for the details which observers should fill in, including measurements and information on local population, market days, and manufactures. The proposal concluded by noting that trials on Blackheath had shown the method to be practical and that the times of rocket and mortar firing would be announced in the papers before they took place.

Whiston's scheme extended Derham's idea of using networks of lay observers and churches as observation posts to the whole country. The list of sponsors of the scheme indicate that it was not just a vain idea of Whiston's but must have been taken seriously. They included Derham himself, mathematician Roger Cotes, instrument-maker and Royal Society demonstrator Francis Hauksbee, publisher and globe-maker John Senex, the clockmaker George Graham and Whiston's friend the Newtonian chaplain Richard Laughton.[39] Whiston began the promised advertisements in the newspapers in the summer of 1715, and these appeared again in 1717.[40] Meanwhile the Northampton surveyor and sundial-maker Edward Laurence gave an enthusiastic description of Whiston's method in his *Young Surveyor's Guide*, the first original text on surveying of the eighteenth century, published in 1716 and seeing two more editions in 1717 and 1736. So the technique was not ignored.[41] Whiston's plans depended on the public to work, since he required lay observers to make accurate measurements and communicate them to him. In 1717 he asked people to observe,

> how plainly, and how many Seconds the [fire]Balls are seen; and if they well can, at what utmost Altitude, and in what Angle from the Meridian

---

39   [Whiston], *Proposals (by Way of Subscription) For a New and most Correct Sett of Maps*, 2.

40   *Post Man and the Historical Account*, July 12, 1715, no. 11150.

41   Edward Laurence, *The Young Surveyor's Guide: Or, a New Introduction to the Whole Art of Surveying Land, both by the Chain and All Instruments Now in Use* (first edition, London, 1716 and third edition, London, 1736), 210–11; on Laurence's book see A.W. Richeson, *English Land Measuring to 1800: Instruments and Practices* (Cambridge, MA: MIT Press, 1966), 150–152.

they are seen: And in what Places within 20 or 30 miles, how many seconds the sound is heard after the first Sight of the Light also.[42]

Cotes expressed scepticism about these expectations in letters to Whiston.[43] Evidently the scheme did not work, as no further mention of it was made by Whiston after 1717. Instead, in 1719 Whiston turned his attention to finding the longitude by magnetic variation and he offered a method using solar eclipses in 1724.

Nevertheless, in 1721, John Senex announced a scheme in the newspapers to make a new map of Surrey using rockets to survey the county.[44] Senex, one of Whiston's subscribers and publisher of his cometary schemes, appealed to the landed interest by claiming the map would be "a specimen for the rest of the counties of England."[45] No more maps followed, but by 1725 Senex was advertising an estate mapping service in collaboration with the surveyor Richard Cushee.[46] Senex presented his Surrey map as full of philosophical and gentlemanly credit. Besides being Whiston's publisher he was a leading globe-maker and surveyor, also printing works for figures such as Halley, Desaguliers, Hauksbee, and Newton. His advertisement explained how,

> A Ball of Fire will be thrown up from the Top of Box-Hill at half an Hour past Eight every Evening for a Fortnight; and Rockets will be let off from proper Eminences near Godalming, Chertsey, and Westram; the first half a quarter of an Hour after the Ball, and the rest at a like interval from each other: Whence such Gentleman as are furnish'd with proper Instruments, will have an Opportunity of determining their own Bearings, and will thereby be enabled to judge of the Justness and Accuracy of the Map when finish'd.[47]

Senex depended on the landed and other interests to secure the project's credit, explaining that he would need the "Countenance and Encouragement

42   *Post Man* 14 May 1717; *Daily Courant* 13 May 1717.

43   Stewart, *Rise of Public Science*, 191. Trinity College Library, MS. R. 4. 42., no. 8, Whiston to Cotes, November 26, 1714; no. 9. Cotes to Whiston, December 2, 1714; no. 11, Whiston to Cotes, April 7, 1715.

44   Laurence Worms, "The Search for John Senex F.R.S.: An Aspect of the Early Eighteenth-Century Book Trade." Paper delivered to the Bibliographical Society 18 January 2000 (accessed 22 July 2015) https://laurenceworms.wordpress.com/2014/01/30/the-search-for-john-senex-f-r-s-an-aspect-of-the-early-eighteenth-century-book-trade/

45   *Post Boy*, August 12–15, 1721, no. 5002; Whiston, *Memoirs*, vol. 1, 191.

46   *Daily Post*, February 10, 1725, no. 1678.

47   *Post Boy*, August 12–15, 1721, no. 5002.

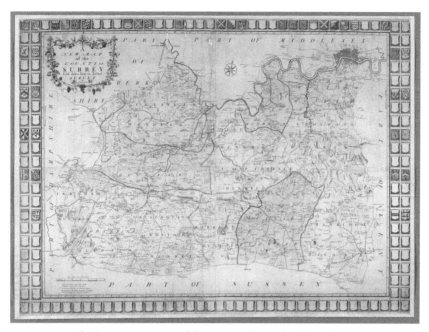

FIGURE 9.1   John Senex, A new map of the county of Surrey laid down from an actual
              survey, 1729, reproduced with permission from the British library board,
              06/07/2021, Cartographic items Maps K.Top.40.5. Note that few of the blank
              shields have been completed with arms.

of the Noblemen and Gentlemen, the Advice and Information of the Virtuoso and Men of Letters, and the Occasional Assistances of Persons in other
Capacities."[48] Subscriptions and receipts were issued from his premises at the
Globe in Fleet Street and other shops. Gentlemen of Surrey were requested to
send in their arms to appear on the map. Although details of the subsequent
works are unknown, Senex completed a four-sheet map, entitled *A New Map
of the County of Surrey Laid Down from an Actual Survey*, with Cushee in 1730
(figure 9.1).[49] The one inch to a mile map displayed north-south minutes of latitude and east-west seconds of time from the London meridian located at the
Royal Observatory. Publication was delayed while Senex awaited arms from
Surrey gentlemen, few of which materialized. Surrey's finest families were perhaps not ready for ingenious surveys, but Senex's project kept Whiston and
Ditton's methods active into the 1730s.

---

48   *Post Boy*, August 12–15, 1721, no. 5002.
49   John Senex, *A New Map of the County of Surrey Laid Down from an Actual Survey* (1729),
      British Library Maps K.Top.40.5; the map was dated 1729, but advertisements indicated it
      was not ready before 1730. See Daily Journal, March 10, 1730; Issue 2862.

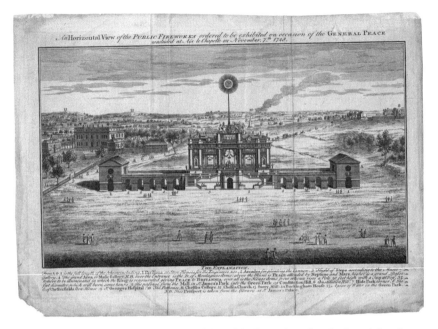

FIGURE 9.2   Anon., An horizontal view of the public fireworks ordered to be exhibited on
occasion of the general peace concluded at Aix la Chapelle on November 7th
1748, copper engraving (London, 1749). Collection of the author.

London again became the centre for rocket experiments in the following
decade. The Green Park fireworks of April 1749 marked the peace of Aix-la-
Chapelle, and their preparations formed the centre of attention for fashionable
London over several months leading up to the display (figure 9.2). In November 1748, an anonymous letter appeared in the *Gentleman's Magazine* proposing a "geometrical use" for the impending display.[50] The author explained that
"rockets... besides the beauty of their appearance, are, or may be, of very great
use in geography, navigation, military affairs, and many other arts... for all kinds
of instantaneous discourse between different stations."[51] The fireworks would
offer an occasion for making more accurate measures of the altitude to which
rockets ascended for the purpose of using them to determine distances from
a signal. Readers living within fifteen to fifty miles of the park were invited
to measure the angle of the rockets at their highest altitude, and the author
noted that Senex had previously experimented with one pound rockets for the

50   Anon., "A Geometrical Use Proposed from the Fire-Works," *Gentleman's Magazine* 18
      (November 1748): 488; Werrett, *Fireworks*, 154–157.
51   "A Geometrical Use... from the Fire-works," 488.

map of Surrey, making them rise to 1400 feet. Soon, other correspondents to the magazine recommended the experiment, refining methods to make the observations.[52]

The author of this letter remained anonymous. It may have been Whiston, who was eighty-one years old at the time. Certainly Benjamin Robins, the Royal Society's expert on ballistics and author of *New Principles in Gunnery* (1742) made observations of the fireworks and reported on them in the *Philosophical Transactions*, and he may have authored the *Gentleman's Magazine* letter. [53] Robins recorded that single rising rockets in the Green Park display reached heights of up to 1800 feet, which suggested rockets might be visible at distances up to 50 miles.[54] With two other fellows, John Ellicot and John Canton, Robins tried further experiments on rocket ascents using pyrotechnics made by a member of the Royal Laboratory at Woolwich, which manufactured fireworks for the crown. When these proved unsatisfactory, a Jewish merchant of Devonshire Square named Samuel Da Costa and an otherwise unknown fireworks maker named Mr. Banks were employed to make rockets which rose to a height of 3,762 feet, double those of the Green Park display.[55] Reporting on the trials, John Ellicott concluded that the best rockets for future observations should be between 2.5 and 3.5 inches diameter, not the highest in altitude, but offering the best balance of cost and degree of ascent.[56] The goal of these experiments remained the same as Whiston and Ditton's, having "great use… in determining the position of distant places, and in giving signals for naval and military purposes."[57]

52   Philo Pyrotechnicus, "Rules for Computing the Height of Rockets, &c.," *Gentleman's Magazine* 18 (December, 1748): supplement, 597; *Gentleman's Magazine* 19 (February, 1749): 55–56.

53   This is the view taken in Werrett, *Fireworks*, 155. Robins published his observations as "Observations on the Height to Which Rockets Ascend," *Philosophical Transactions of the Royal Society* 46 (1749): 131–133.

54   Robins, "Observations," 132–3.

55   John Ellicot, "An Account of Some Experiments, Made by Benjamin Robins Esq; F. R. S. Mr. Samuel Da Costa, and Several Other Gentlemen, in Order to Discover the Height to Which Rockets May Be Made to Ascend, and to What Distance Their Light May be Seen," *Philosophical Transactions of the Royal Society*, 1749, 46: 578–584; the trials were also noticed in *The Family Memoirs of the Rev. William Stukeley, M.D., and the Antiquarian and Other Correspondence of William Stukeley, Roger & Samuel Gale*, etc., 3 vols. (Publications of the Surtees Society, vol. 76; Durham; London, 1882–87), vol. 2 (Durham; London, 1883), 374. Recall that Whiston said shells could be fired from mortars up to 1000 feet high, and if something could be fired 6400 feet high, it would be seen from 100 miles distance.

56   Ellicot, "An Account," 582.

57   Benjamin Robins, *Mathematical Tracts of the Late Benjamin Robins* (London, 1761), 323.

The use of rockets as signals proliferated after these experiments. Even before Robins's experiments, schemes to use rockets for military communication were proposed. In 1745 one William Gee suggested a means to control Jacobite rebels in Scotland using "an early Intelligence from any part of the Kingdom by Signals; I think Rockets might be the best... by placing Signals at proper distances, the whole Country would in one Hour, be upon their Guard." The scheme recalled the range of Whiston's project, since "a Signal in the Evening, might be made, to Reach, one Hundred Miles, provided the signal makers were careful to look out, and to be in readiness."[58] The scheme did not materialize, but signal rockets were manufactured in the Royal Laboratory at Woolwich in subsequent decades, and were used frequently around the world by the British army and navy in the second half of the eighteenth century.[59] Daniel Defoe described the use of rockets as naval signals in his fictional *A New Voyage Round the World* of 1725, but the idea first appeared in print in a naval publication with the English translation of Sebastien Francois Vicomte Bigot De Morogues's *Tactique Navale*, published in 1767.[60] In 1799, Captain James Gambier of the Royal Navy published *Night Signals and Instructions for the Conduct of Ships of War*, introducing a standardized system which included using flights of rockets to indicate exercises and messages. A contemporary report explained that,

> A code of night signals has been lately established, and sent down by the Admiralty to the several posts on the heights round the coast, and a cordon of repeating frigates are stationed at convenient distances, between whom and the shore a ready communication can at all times be kept up by rockets and variegated lamps.[61]

---

58 Proposal of William Gee to [Newcastle] for a system of signalling by rockets in the evening to be instituted, to be used if the army in the north required assistance against the rebels from other parts of the country or when an enemy invasion in remote parts seemed imminent. October 9th 1745. National Archives, Kew, SP 36/71.

59 John Maskall described signal rocket manufacture in the manuscript book kept at the Laboratory, "Artificial Fireworks," 3 vols. (1785), vol. 1, ff. 73–5, Getty Research Institute, Los Angeles; see also Captain Robert Jones, *A New Treatise on Artificial Fireworks* (London, 1765), 135–6.

60 Sebastien Francois Vicomte Bigot De Morogues, *Tactique navale* (Paris, 1763); Sebastien Francois Vicomte Bigot De Morogues, *Naval Tactics: Or a Treatise of Evolutions and Signals* (London, 1767), 51–53; for an earlier scheme of signal lights using lanterns made with oil, pitch and tallow, see Captain Robert Hamblin, "Signal Lights", Patent no. 517, dated July 21, 1730, British Library.

61 *The Times*, September 5, 1804.

The use of rockets to determine position on land also continued through the remainder of the eighteenth century. In 1750 the third edition of *The Practical Surveyor* by Samuel Cunn (writing under the name of his friend's clerk John Hammond) noted the value of using rockets as signals for surveys between land and sea, whereby a rocket fired from a boat could signal its position.[62] Natural philosophers also made use of rockets to determine longitude. In June 1769, the astronomer royal Nevil Maskelyne and John Canton, who had worked with Robins on rocket ascents two decades previously, used rockets to measure the longitude of Spital Square in east London during preparations to view the Transit of Venus.[63] When the astronomer Francis Wollaston made meridian observations from the parsonage of Chislehurst in south London beginning in the 1770s, he also measured its longitude using rockets. This was done in conjunction with measures using triangulation to produce a mean of several methods and measures, so as to ensure accuracy.[64] Such a method reflected maritime practice in which chronometry and astronomical observations would both be used to find longitude at sea. Wollaston's rockets were fired from Loampit Hill, Deptford, location of Alexander Aubert's private observatory, and observed by Wollaston in Chislehurst, by Nevil Maskelyne in Greenwich, John Ellicott in Horseley Lane and the physician William Heberden in Pall Mall, London. Scots mathematician John Playfair, who also took part in the experiments, claimed they were a great success. The difference in longitude was determined between these points with great precision, to "a fraction of a second."[65]

A decade later, between 1787 and 1789, the Scottish surveyor and antiquary William Roy undertook surveys to measure the difference in longitude between

62   John Hammond, *The Practical Surveyor: Containing the Most Approved Methods for Surveying of Lands and Waters*, third edition (London, 1750), 120; on the attribution to Cunn, see Richeson, *English Land Measuring*, 150.

63   John Canton, "A Letter to the Astronomer Royal, from John Canton, M.A., F.R.S. Containing His Observations of the Transit of Venus, June 3, 1769, and of the Eclipse of the Sun the Next Morning," *Philosophical Transactions* 59 (1769): 192–194, on 193.

64   Francis Wollaston, "On a Method of Describing the Relative Positions and Magnitudes of the Fixed Stars; Together with Some Astronomical Observations," *Philosophical Transactions* 74 (1784): 181–200, on 183; Francis Wollaston, *Fasciculus Astronomicus: Containing Observations of the Northern Circumpolar Region; Together with Some Account of the Instrument with which they were Made and a New set of Tables by which they were Reduced to the Mean Position for the Beginning of January 1800* (London, 1800), 6.

65   Anon. [John Playfair], review of Baron Zach, "L'attraction des montagnes et ses effets sur les instruments d'astronomie constatés et déterminés par des observations... fait en 1810... par le Baron de Zach, Avignon, 1814," *Bibliothèque britannique, ou Recueil extrait des ouvrages anglais* 57 (Geneva, 1814): 3–28, on 15; the review is attributed to John Playfair as "Review of Baron de Zach, Attraction des Montagnes," in *The Works of John Playfair*, vol. 4 (Edinburgh, 1822) 467–500, 485–486.

the observatories of Paris and Greenwich, after Cassini de Thury informed the British that there was a difference in results made by the observatories of some ten seconds of arc.[66] Sven Widmalm has proposed that Roy helped to introduce a new French style of surveying practice into Britain at this time. Roy's choice of signals bears this out. Roy's surveys followed Cassini de Thury in triangulating meridian arcs to provide the basis for cartography, and used precision instruments, especially a new theodolite by Jesse Ramsden, which helped increase accuracy levels from about two seconds to half a second in angular measurements. Triangulation involved measuring a base line exactly on a flat plane with chains laid end to end. A surveyor placed the theodolite at one end and trained it on a surveyor holding a flagstaff on top of a landmark. The angle between one end of the base line and the landmark was noted, and then the angle between the landmark and the other end of the base line. This formed a triangle in which two angles and one side were known, allowing calculation of the length of the other sides and so the distance from the baseline to the landmark. Then the new line from the baseline to the landmark could serve as a new baseline for constructing the next triangle. In this system, illuminated signals were only needed when it was difficult to see the landmark. When Cassini and Roy sought to triangulate across the channel, foggy weather led them to choose brilliant white lights as signals. These consisted of "Bengal" or "blue lights" which the British had recently encountered in India, made with a mixture of nitre, sulphur and trisulphate of arsenic which produced a bright blue-white flame when the mixture was burnt in copper cups. They were placed in vertical rows on flagstaffs, and were visible up to forty miles away.[67]

In a report on the survey, Roy noted that there was another way to determine the difference in longitude between the observatories: "The astronomical difference of time may... be obtained by experiments on the instantaneous explosion of light; but these I would propose to be made subsequently to the

---

66   Nevil Maskelyne, "Concerning the Latitude and Longitude of the Royal Observatory at Greenwich; With Remarks on a Memorial of the Late M. Cassini de Thury," *Philosophical Transactions* 77 (1787): 151–187; Sven Widmalm, "Accuracy, Rhetoric, and Technology: The Paris-Greenwich Triangulation, 1784–88," in Tore Frängsmyr, John Heilbron, and Robin E. Rider, eds., *The Quantifying Spirit in the Eighteenth Century* (Berkeley: University of California Press, 1990), 179–206.

67   J.P. Martin and Anita McConnell, "Joining the Observatories of Paris and Greenwich," *Notes and Records of the Royal Society* 62 (2008): 355–372, on 363, 372 n. 24; Rachel Hewitt, *Map of a Nation: A Biography of the Ordnance Survey* (Granta, 2011), 85–6; The *Encyclopédie methodique*, "Physique" par MM. Monge, Cassini, Bertholon, Hassenfratz, &c. &c. (Paris, 1819) vol. 3, 161, describes "Feu Blanc Indien. White fire, very brilliant, which is distinguishable at a very great distance."

trigonometrical operations."[68] Roy announced that such a method had a secondary place in the survey, used to verify the more reliable method of triangulation. Nevertheless, astronomers with clocks and transit instruments might
observe alternating flashes of gunpowder and Bengal lights at appropriate stations. Roy reckoned the method would yield a time difference between the
two locations "to a very considerable degree of accuracy, and probably more
to be relied upon than that resulting from the comparison of the observations of the heavenly bodies."[69] Roy was dismissive, however, of using rockets,
since they generated "too small a body of white light" to be visible at long distances.[70] Alternatively, Roy proposed the use of balloons carrying explosive
charges, which would be raised to a given height, held in position by a string,
and exploded using a fuse. One explosion would light the fuse for another, and
others would follow at regular time intervals. Even then, Roy reckoned that the
explosive method would need to be repeated many times to match the accuracy that "may infallibly be brought by trigonometry."[71]

Roy's reluctance to use rockets and preference for exploding flashes of gunpowder can be clearly situated in a French tradition dating back to the 1730s,
whose difference from British activities suggests the enduring local influence
of Whiston and Ditton's scheme in Britain.[72] In 1735 Charles-Marie de la Con-
damine had suggested using a cannon explosion in measures of an arc of the
meridian and in the following decade members of the Paris Academy of Sciences experimented in the use of gunpowder flashes for surveying purposes.
In 1739 Cassini de Thury and Nicolas-Louise Lacaille used the flashes of ten-
pound gunpowder charges to determine the difference in longitude between
the Hermitage of St. Clair near Sete, and Mount St. Victoire near Aix as part
of their national survey of France.[73] This was no doubt the source of Roy's
approach, while the use of balloons also suggests a French influence.

---

68   William Roy, "An Account of the Mode Proposed to be Followed in Determining the
      Relative Situation of the Royal Observatories of Greenwich and Paris. By Major-General
      William Roy, F.R.S. and A.S." *Philosophical Transactions* 77 (1787): 188–469, on 214.
69   Roy, "An Account of the Mode Proposed," 215.
70   Roy, "An Account of the Mode Proposed," 215.
71   Roy, "An Account of the Mode Proposed," 216.
72   The history of French uses of 'fire signals' in determining longitude was traced in an essay
      by the Chevalier Bonne, "De la détermination des longitudes terrestres par le moyen des
      signaux de feu," *Mémorial du Dépôt général de la guerre* 3 (1826): 25–60.
73   *Journal du voyage fait par ordre du Roy à l'équateur: servant d'introduction historique a la
      mesure des trois premiers degrés du méridien*, volume 2 (Paris, 1752), 47; J. E. Portlock, *Mem-
      oir of the Life of Major-General Colby, Together with a Sketch of the Origin and Progress of
      the Ordnance Survey of Great Britain and Ireland* (London, 1869), 86; Bonne, "De la déter-
      mination des longitudes," 39–40.

The practice of using gunpowder flashes rather than rockets for surveys continued into the first decades of the nineteenth century.[74] But rockets were used again, by both the British and the French. In 1820, the adventurer and naval officer William Fitzwilliam Owen wrote to the Board of Longitude to propose a scheme for measuring longitude which bore a remarkable resemblance to Whiston's earlier scheme, though Owen claimed it as his own.[75] Five years later John Herschel and Edward Sabine co-operated with the Frenchmen Largeteau and Bonne to make a new determination of the difference in longitude between the Paris and Greenwich observatories, this time by firing rockets supplied by the French at a series of stations between Greenwich and Paris. The time differences between observers could be found by simultaneous observation of the rockets and the longitude worked out. Despite losing much of the data, Herschel concluded a difference of nine degrees, twenty-one minutes and six seconds.[76] In 1839, Thomas Romney Robinson of the Armagh Observatory measured the difference of longitude of the observatories of Armagh and Dublin using rockets and pyrotechnic signals continued to be used in surveys and longitude determinations thereafter.[77]

3   Conclusion

In 1840, J.M.W. Turner exhibited a painting at the Royal Academy depicting *Rockets and Blue Lights (Close at Hand) to Warn Steam Boats of Shoal Water* (figure 9.3). In the scene, spectators watch from a beach as the brilliant light of rockets and flares explode in the foggy atmosphere over the sea, warning of shallow or "shoal" waters. Early critics attacked the painting for its ambiguity,

---

74   See e.g. Joseph Nicollet, "Mémoire sur la mesure d'un arc du parallèle moyen entre le pôle et l'équateur; par M. Brousseaud, colonel an Corps royal des ingénieurs-géographes, et Nicollet, astronome adjoint au Bureau des Longitudes," *Connaissance des temps ou des mouvements célestes* (1829): 252–295.

75   See the letter from Captain W.F. Owen to Mr. J.W. Croker, 20 May 1820, Papers of the Board of Longitude, RGO 14/51:212r. Accessed online 23 July 2015, << http://cudl.lib.cam.ac.uk/view/MS-RGO-00014-00051/432>>

76   John Herschel, "Account of a Series of Observations made in the summer of the year 1825, for the purpose of determining the difference of meridians of the Royal Observatories of Greenwich and Paris," *Philosophical Transactions* 116 (1826): 77–126; I am indebted to Michael Kershaw for providing me with a survey of this work, from his forthcoming book *Where is Paris? Longitude on Land.*

77   Rev. Thomas Romney Robinson, "On the Differences of Longitude Between the Observatories of Armagh and Dublin, Determined by Rocket Signals," *Transactions of the Royal Irish Academy* 19, part 1 (1841): 121–146.

FIGURE 9.3   Joseph Mallord William Turner, *Rockets and Blue Lights* (*Close at Hand*) *to Warn Steam Boats of Shoal Water*, exhibited 1840. Courtesy of the Clark Art Institute. clarkart.edu

"we were totally at a loss to know what it meant… A thing more without form or shape of anything intended we never saw."[78]

Turner's painting presents a triangulation of land, sea, and pyrotechnics that has also been the focus of this essay. Like the painting, William Whiston and Humphrey Ditton's longitude scheme using rockets was maligned by contemporaries, ridiculed as the misguided labour of a madman. But as this essay has shown, Whiston did not intend the project as applicable only at sea, and once deployed on land, his method proved more successful. The use of rockets as signals was so ubiquitous subsequently as to make the originality of this function invisible. But Whiston and others innovated by imagining fireworks displayed for state occasions as potential instruments of measurement. In the first half of the eighteenth century, natural philosophers and artisan collaborators busily explored the capacities of rockets as signals for surveying and warning. After Whiston's investigations, John Senex, Benjamin Robins, and others continued the experiments, until rockets and later other forms of pyrotechnics

---

78   Anon., "Royal Academy Exhibition, &c." *Blackwood's Edinburgh Magazine* 48 (1840): 374–386, on 384.

came to used routinely for position-finding and surveying. By the time Turner's painting was exhibited such arrangements were quite familiar. A textbook on astronomy noted in 1841 that the longitude could best be determined by "the explosion of rockets... These are almost instantaneous occurrences, and hence longitudes may be obtained from them to a higher degree of accuracy than by any other method."[79]

The Whiston-Ditton longitude solution depended on landed resources and simultaneously generated territorial knowledge. It was not rolling ships at sea which determined the fate of Whiston's scheme, but the rolling hills of the Surrey countryside, where enterprising experimenters set off rockets to support their ingenious cartographic schemes. Appreciating the significance of scientific methods, techniques, and ideas thus appears to depend significantly on the spatial framing within which they are examined. As a maritime venture, studied in relation to longitude at sea, the historical trajectory of Whiston's rockets quickly peters out. But reframed in a context of land surveying, their career is seen to be more enduring.

---

79   Ebenezer Porter Mason, *Introduction to Practical Astronomy: Designed as a Supplement to Olmsted's Astronomy* (New York, 1841), 124.

Index